Multiscale Modeling of Degradation in Lithium-Ion Batteries

Von der Fakultät für Maschinenbau
der Technischen Universität Carolo-Wilhelmina zu Braunschweig

zur Erlangung der Würde

eines Doktor-Ingenieurs (Dr.-Ing.)

genehmigte Dissertation

von: Dipl.-Ing. Fridolin Röder
geboren in: Malsch

eingereicht am: 4.9.2018
mündliche Prüfung am: 18.1.2019

Vorsitz:

Prof. Dr.-Ing. Carsten Schilde

Gutachter:

Prof. Dr.-Ing. Ulrike Krewer
Prof. Dr. Richard D. Braatz

2019

Bibliographic information published by the Deutsche Nationalbibliothek

The Deutsche Nationalbibliothek lists this publication in the Deutsche Nationalbibliografie; detailed bibliographic data are available on the Internet at http://dnb.d-nb.de .

ISBN 978-3-8325-4927-5

Logos Verlag Berlin GmbH
Comeniushof, Gubener Str. 47,
10243 Berlin
Tel.: +49 (0)30 42 85 10 90
Fax: +49 (0)30 42 85 10 92
INTERNET: https://www.logos-verlag.de

Acknowledgments

The research of this thesis was conducted during the last five years at the Institute of Energy and Systems Engineering (InES) at the TU Braunschweig and at the Massachusetts Institute of Technology (MIT).

First, I would like to thank my supervisor Prof. Dr.-Ing. Ulrike Krewer for giving me the opportunity to work on this interesting and challenging topic. She was always unprejudiced for unorthodox and novel solutions and at the same time helped me to shape my ideas with critical discussions and a remarkable persistence to all the necessary details. Further, I would like to thank Prof. Richard D. Braatz. from MIT for the fruitful discussions, which often turned problems into new opportunities.

Many thanks to my colleagues at the InES and at the Battery LabFactory Braunschweig (BLB), for many interesting and inspiring scientific and non-scientific discussions. Especially, I would like to thank the members of the battery group Angelica Staeck, Nan Lin, Lars Bläubaum, Fethi Belkhir, Oke Schmidt, Patrick Schön, and Walter Cistjakov. I would also like to thank the technical staff at InES, especially Wilfried Janßen for his support with IT issues, and Horst Müller and Ina Schunke for their enormous helpfulness with any administrative issue at the institute. Moreover, I would like to thank the students/coworkers Joseph Sullivan, Vijayshankar Dandapani, Florian Baakes, and Niels Brinkmeier for critical review and proofreading of the manuscript of this thesis and Daniel Schröder, Christine Novak, Sören Sonntag, Tunay Okumus, Vincent Laue, Georg Lenze, Tom Patrick Heins, Marco Heinrich, Nina Harting, and Nicolas Wolff for scientific discussions, joint research and publications.

Finally, I would like to thank my friends and family, who supported and encouraged me most during this period.

Contents

List of Figures

List of Tables

List of Symbols

Latin letters

- a activity, –
- a_s specific surface area, m^{-1}
- A pre exponential factor, s^{-1}
- A surface area, m^2
- $\bar{A}$ amplitude, $\mathrm{A\ m}^{-2}$
- A_e^{RK} Redlich Kister coefficient, –
- b fragments through cracking, m^{-1}
- B birth rate, $\mathrm{s}^{-1}\ \mathrm{m}^{-4}$
- c concentration, $\mathrm{mol\ m}^{-3}$
- C^0 standard state concentration, $\mathrm{mol\ m}^{-3}$
- C^{DL} double layer capacitance, $\mathrm{F\ m}^{-2}$
- C_{theo}^{As} theoretical capacity, $\mathrm{A\ s\ m}^{-2}$
- d^{film} film thickness, m
- $\hat{d}^{\mathrm{film}}$ local film thickness, m
- D death rate, $\mathrm{s}^{-1}\ \mathrm{m}^{-4}$
- D diffusion coefficient, $\mathrm{m}^2\ \mathrm{s}^{-1}$
- E estimate of the kMC output parameter
- E electrode potential, V
- $\hat{E}^A$ specific activation energy, $\mathrm{J\ mol}^{-1}\ \mathrm{m}^{-1}$
- f frequency, s^{-1}
- f density function, m^{-x}

- F Faraday constant, A s mol^{-1}
- h fraction density, m^{-1}
- i_0 exchange current density, A m^{-2}
- I current density, A m^{-2}
- j current density, A m^{-3}
- J process, –
- J surface current density, A m^{-2}
- k shape parameter, –
- k_{ct} exchange current density rate constant, m$^{2.5}$ mol$^{-0.5}$ s^{-1}
- k^f forward reaction rate constant, [s,mol,m]
- k^b backward reaction rate constant, [s,mol,m]
- K_P proportional factor, –
- K_I integral factor, –
- L length, m
- L lattice site, –
- r radial coordinate, m
- M mean value of kMC output parameter
- n number (of), –
- N_s site density, m^{-2}
- o_s site occupancy number, mol^{-1}
- p kMC output parameter
- p probability, –
- P second particle radius, m
- q reaction flux, mol m^{-2} s^{-1}
- Q source flux, mol m^{-2} s^{-1}
- r_s roughness factor, –
- R ideal gas constant, J mol^{-1} K^{-1}
- R_s particle radius, m

- R electrical resistivity, Ω m
- S standard deviation of kMC output parameter
- t time, s
- t_p transference number, –
- T temperature, K
- V volume, m^3
- x coordinate, m
- x^{Li} interacalation fraction, –

Greek letters

- α^{cr} cracking kernal, $(\mathrm{runtime})^{-1}$
- β symmetry factor, –
- β^{agl} agglomeration kernal, m^3 $(\mathrm{runtime})^{-1}$
- δ thickness, m
- Γ microscopic rate, s^{-1}
- $\Delta\Phi$ difference in electrical potential, V
- ΔG^0 standard state Gibbs free energy, J mol^{-1}
- ΔL size of surface site, m
- ΔR radius difference / m
- Δt difference in time, s
- ϵ volume fraction, –
- ζ uniform distributed random number, –
- θ surface fraction, –
- ϑ surface site state, –
- $\vartheta^{\mathrm{film}}$ film site state, –
- $\vartheta^{\mathrm{lattice}}$ lattice site state, –
- κ smoothing factor, –
- λ tolerance factor, –

- λ^{kMC} kMC output parameter
- λ scale parameter, µm^{-1}
- μ^0 standard state chemical potential, J mol^{-1}
- ν stoichiometric factor, –
- σ electrical conductivity, S m^{-1}
- σ_{De} diffusional ionic transport coefficient, V S m^{-1}
- τ tortuosity, –
- Φ electrical potential, V
- Ψ boolean indicating selected process, –

Indexes

- e iteration for estimation correction, –
- i kMC times step, –
- j microscopic process, –
- l lattice site, –
- m species, –
- s sequence, –
- v parallel instance, –
- ι last kMC time step within sequence
- ς sequence, –

List of Abbreviations

- ads – adsorption site
- area – surface area
- ct – charge transfer
- CB – carbon black
- CFD – computational fluid dynamics
- CPU – central processing unit
- DFT – density functional theory
- e – electrolyte phase
- eff – effective
- EC – ethylene carbonate
- film – film phase
- G – graphite
- kMC – kinetic Monte Carlo method
- LC – lithium carbonate
- LEDC – lithium ethylene di-carbonate
- max – maximal
- MD – molecular dynamics
- MP – multiparadigm
- MPA – multiparadigm algorithm
- num – number
- NMC – nickel manganese cobalt
- ODE – ordinary differential equation
- par – parallel

- P2D – pseudo two dimensional
- PDE – partial differential equation
- PSD – particle size distribution
- PVDF – polyvinylidenfluorid
- QSSA – quasi stationary approximation
- s – solid phase
- seq – sequence
- S – solvent
- SP – single particle
- SEI – solid electrolyte interface
- SMG – surface modified graphite
- tot – total
- vol – volume
- VC – vinyl chloride

Abstract

The objective of this thesis is the development of methods for multiscale simulation of degradation in lithium-ion batteries. There are interactions between operational processes, side reactions, film growth and electrode restructuring throughout the scales, i.e. electrode scale, particle scale, mesoscale, and atomistic scale. However, those aspects are commonly modeled independently and on single scales. In this thesis, a novel multiscale methodology has been developed, which bridges a wide range length scales, and enables to study those interactions. The thesis provides algorithms, models, and concepts and applies them for analysis of multiscale effects in lithium-ion batteries.

First, concept and algorithms for direct coupling of continuum and kinetic Monte Carlo (kMC) models were developed. These methods were applied to develop multiscale models for simulation of heterogeneous surface film growth problems in lithium-ion batteries. Further, electrode heterogeneity and electrode restructuring is modeled using statistically distributed parameters and population balance equations. With this, the thesis provides a comprehensive methodology to study multiscale interactions in batteries.

The thesis provides a guideline for selection and tuning of algorithms for direct coupling of kMC and continuum models by systematically comparing accuracy and computational cost. This enables to design a multiscale model for analysis of complex degradation mechanisms at electrochemical interfaces. One problem of this class is the growth of the, so-called, solid electrolyte interface (SEI) at negative electrodes in lithium-ion batteries. This phenomenon has been modeled and analyzed in detail. Simulation results could reproduce experimental observation from literature as well as electrochemical experiments. Among others, this includes: the slope of potential during the first charge, film compositions, and impact of particle size on reversible capacity. Distinct multiscale effects during film growth process have been revealed. Results indicate that the structure of the surface film considerably depends on the charging protocol. Thus, consideration of multiscale interaction enables to optimize charging protocols to achieve desired surface film structures. Moreover, macroscopic heterogeneity impacts battery performance and can trigger degradation of surface films or restructuring of electrodes, which has been evaluated with the multiscale methodology in detail.

To sum up, this thesis gives significant new physical insight into degradation processes and multiscale interaction, as processes from the electrode to the molecular scale have been connected. Moreover, the work provides a large set of novel algorithms, models and methods for multiscale analysis of lithium-ion batteries. This opens new opportunities of research and development in this and related fields. Application of this comprehensive approach will enable the optimization of batteries, their materials, and production procedures. This will increase life time, ensure safety, and reduce cost of lithium-ion batteries, and thus significantly contributes to transfer the energy market towards renewable energies and in particular to the development towards electromobility.

Kurzfassung

Das Ziel dieser Dissertation ist die Entwicklung einer Methode zur Multiskalenmodellierung der Degradation von Lithium-Ionen-Batterien. Prozesse während des normalen Betriebs sowie Prozesse aufgrund von Nebenreaktion interagieren signifikant untereinander und werden jeweils durch Degradationsprozesse wie etwa der Restrukturierung der Elektroden oder dem Wachstum von Oberflächenschichten beeinflusst. Dies führt zu einer starken Interaktion über viele Größenskalen, also auf Elektrodenebene, Partikelebene, Mesoebene oder atomistischer Ebene. Oft werden diese Aspekte nur getrennt und auf einer begrenzten Größenskala betrachtet. In dieser Dissertation wurde eine neue umfassende Methode zur Multiskalenmodellierung entwickelt, welche die Überbrückung eines sehr großen Skalenbereichs ermöglicht, um diese Interaktionen zu analysieren. Die Arbeit stellt dabei Algorithmen, Modelle und Konzepte vor und wendet diese für die Analyse von Multiskaleneffekten in Lithium-Ionen-Batterien an.

Zunächst wurden Konzepte und Algorithmen zur direkten Kopplung von Kontinuumsmodellen und Modellen auf Basis der kinetischen Monte Carlo (kMC) Methode entwickelt. Diese wurden dann bei der Entwicklung eines Multiskalenmodells zur Simulation von heterogenen Schichtwachstumsprozessen in Lithium-Ionen-Batterien angewandt. Außerdem konnte die Heterogenität auf Elektrodenebne sowie die Elektrodenrestrukturierung durch Anwendung von statistisch verteilten Parametern sowie Populationsbilanzen abgebildet werden. Hiermit stellt die Arbeit eine umfassende Methodik zur Untersuchung von Multiskaleninteraktion in Batterien bereit.

Die Dissertation stellt eine Richtlinie zur Auswahl und Konfiguration von Algorithmen zur Kopplung von kMC und Kontinuumsmodellen dar, da diese systematisch auf Genauigkeit und Rechenzeit analysiert wurden. Dies ermöglicht den Entwurf von Multiskalenmodellen zur Analyse von komplexen Degradationsmechanismen an elektrochemisch aktiven Oberflächen. Ein Beispiel aus dieser Problemklasse ist das Wachstum der so genannten 'solid electrolyte interface' (SEI) an negativen Elektroden in Lithium-Ionen-Batterien. Dieses Phänomen wurde im Detail modelliert und analysiert. Die Simulationsergebnisse konnten dabei experimentelle Beobachtungen aus der Literatur und aus elektrochemischen Messungen reproduzieren. Dazu zählen unter anderem folgende Aspekte: Verlauf des Potentials beim ersten Laden, typische SEI Zusammensetzungen, Einfluss der Partikelgröße auf die reversible Kapazität. Außerdem konnte klare Multiskaleninteraktionen beim Schichtwachstumsprozess herausgearbeitet werden. Die Ergebnisse weisen darauf hin, dass die Struktur der Oberflächenschicht signifikant vom Ladeprotokoll abhängt. Die Berücksichtigung von Multiskaleneffekten ermöglicht also die Optimierung von Ladeprotokollen, um bestimmte Schichtstrukturen zu realisieren. Hinzu kommt, dass die Heterogenität von makroskopische Strukturen die Batterieperformance beeinflusst und Degradationsprozesse, wie etwa Schichtwachstum oder Elektrodenrestrukturierung, auslösen kann. Diese Aspekte konnten ebenfalls mit Hilfe der Multiskalenmodellierung im Detail evaluiert werden.

Zusammenfassend enthält diese Arbeit wesentliche neue physikalische Erkenntnisse zu Degradationsprozessen sowie zur Multiskaleninteraktion. Dies wurde durch Kopplung von Prozessen von Elektrodenebene bis zur molekularen Ebene ermöglicht. Hinzukommt, dass durch diese Arbeit ein breites Repertoire an neuen Algorithmen, Methoden und Modellen zur Multiskalenanalyse von Lithium-Ionen-Batterien bereitgestellt wird. Die Anwendung dieses umfassenden Ansatzes wird die Optimierung Batterien, ihrer Materialien sowie der Produktionsprozeduren ermöglichen. Hierdurch können zukünftig höhere Lebenszeiten, höhere Sicherheit sowie reduzierte Produktionskosten erreicht werden. Somit trägt die Arbeit signifikant zur Energiewende und insbesondere zur Entwicklung der Elektromobilität bei.

Chapter 1

Motivation and Scope

1.1 From single atoms to advanced energy storage systems

Designing technical products based on fundamental physical knowledge is a vision which drove and will drive engineers in their research and development. Astonishingly successful theories have been developed by physicists, which enable the description of atomistic behavior from a first principle, i.e. without experimental input. Since technical systems are composed of atoms and their interactions can already be simulated, the vision of designing systems from the atomistic level up truly seems achievable. But even the simplest technical systems currently cannot be designed from a pure first principle approach. The consideration of millions and millions of atoms and their interaction is presently far beyond the capability of our computational equipment. Nevertheless, it is worth bringing this vision forward and trying to bridge the scales in order to gain deeper understanding of our complex technical systems in the future.

In this work, lithium-ion batteries, which are an electrochemical energy storage system, are studied. Lithium-ion batteries are rechargeable, and their energy is converted between electrical and chemical energy through electrochemical reactions. Advanced materials, serving as a lithium host structure, enable a very high cycle stability as well as the capability for highly dynamic load. Due to their remarkable properties, they drew a large amount of attention during the last years, and have a major role in enabling the energy market to move from fossil fuels to renewable energy sources and, in particular, to electromobility. Even though lithium-ion batteries are already widely in use, a sufficiently detailed physically based description of their performance, degradation and abuse behavior is still not available. Further, there are many materials considered to be materials of the next generation, which need to be understood. To enable analysis and optimal operation of batteries, there is a need for new and more accurate models.

Physical models of lithium-ion batteries aim to describe these systems and their degradation based on fundamental physical knowledge. Many physical theories and simulation techniques are available. These techniques vary in accuracy and computational cost for numerical solution. Visionary models would cover the full range from fundamental physical theory up to the observed phenomena, e.g. performance, ageing and abuse, while providing the infinitesimally small resolution of an actual technical cell. This is of course far beyond the state of the art. Nevertheless, depending on the actual application, it is possible to realize adequate physical models, which consider certain multiscale effects. Thus, the aim of this thesis is to develop models and strategies adequate for detailed simulation of certain multiscale effects for degradation of lithium-ion batteries, in order to contribute to this vision.

In the following, the structure, function, and degradation of lithium-ion batteries is outlined and available physical models and their paradigms are summarized. With this in mind, the scope of the thesis is given.

1.2 Structure, function, and degradation of lithium-ion batteries [1]

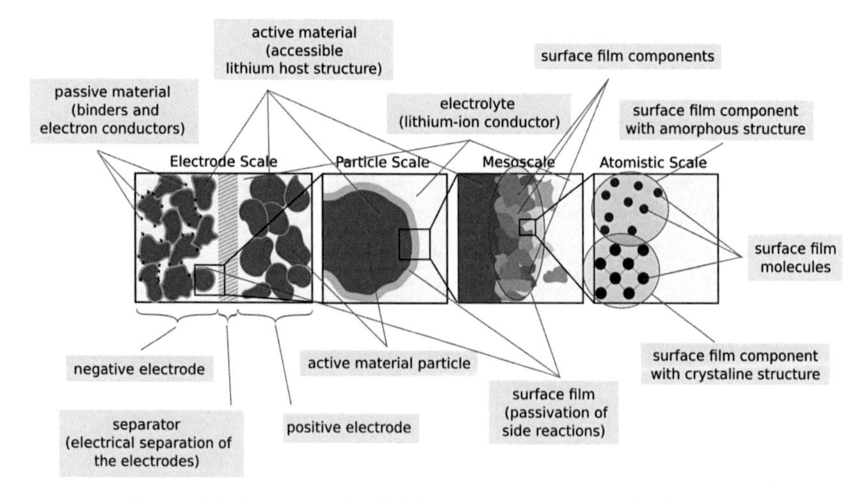

Figure 1.1: Structure of a lithium-ion battery on multiple scales.

In Figure 1.1 the structure of a lithium-ion battery is illustrated. A lithium-ion

[1]Part of this section has been published in (Röder et al., Batteries and Supercaps, 2:248–265, 2019 [1]) and is reproduced with permission from Wiley-VCH Verlag GmbH & Co. KGaA

battery cell is composed of a negative and a positive electrode, which are electrically separated by a separator. The electrodes are electrically connected to the current collector foils at their sides opposite to the separator. The electrodes of a battery are porous and contain active and passive material. The active material, e.g. graphite or nickel manganese cobalt (NMC), provides the accessible host structure for lithium. Passive materials are included for the binding of the materials and electron conduction. The pores of the electrodes and the separator is filled with electrolyte, which serves as lithium-ion conductor. The electrolyte is a lithium salt, e.g. $LiPF_6$, dissolved in a solvent, e.g. ethylene carbonate (EC). Usually, the electrochemical stability window of the electrolyte is smaller than the electrical potential between negative and positive electrodes. This leads to the decomposition of the electrolyte components at the negative electrode. The decomposition products form a solid film on the surface of the active material particles, which is composed of various amorphous and crystalline components. This surface film functions as a lithium-ion conductor and is called the solid electrolyte interface (SEI). The physical properties and geometrical arrangement of the battery components introduced above determine the properties and degradation of the system. During operation of a battery, electrons are conducted from one

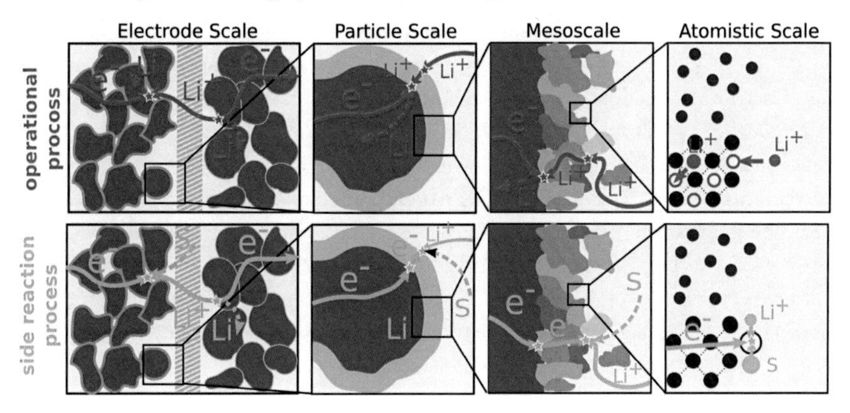

Figure 1.2: Illustration of processes within a lithium-ion battery during charging, with lithium (Li), lithium-ions (Li^+), electrons (e^-), and solvent components (S).

electrode to the other through an external electrical circuit. In Figure 1.2 (top) the charging process of a battery is shown. Electrons enter the negative electrode and exit the positive electrode. At the negative electrode, electrons are conducted through the solid to the active material particle surface, where they electrochemically react at the active material/surface film interface with lithium-ions to form lithium. The lithium then diffuses into the active material host structure. Lithium-ions are produced at the positive electrode and transported through the electrolyte and the surface film. Overall, with this process, lithium is transferred from the positive electrode host structure to the negative electrode host structure during charging and vise versa during discharging.

In Figure 1.2 (bottom) the side reaction process is illustrated. As can be seen, the process is very similar to the processes during normal operation. However, instead of inserting lithium in the negative electrode host structure, lithium electrochemically reacts with the solvent components and forms a surface film, i.e. the SEI. Since the electrochemical reaction here takes place at the surface film/electrolyte interface, the limitations of the process are different from those of the process during normal operation. The process could be limited by electron conduction of the surface film or solvent diffusion to the interface. In general, the operational process and side reactions take place in parallel. However, side reactions are strongly limited by the surface film, and thus their rates significantly decrease with increasing film thickness. A special situation arises at the first charge of a lithium-ion battery, because no film has been formed yet. Thus, at the first charge of the battery, the side reactions are the dominant reaction process at the negative electrodes.

As can be seen in Figure 1.2, for both processes, various aspects at different length scales are of importance. On the electrode scale, solvent diffusion and lithium-ion and electron conduction within the porous electrodes cause differences in electrical potentials and concentrations, which determine reaction rates at the interfaces. On the particle scale, diffusion of lithium into and out of the host structure leads to a concentration gradient inside the active material particles, which depends on the diffusion pathways in particles. On the mesoscale, lithium-ion and electron transport through the surface film depend on the particular structure and composition of the surface film. On the atomistic scale, diffusion, phase change, and reaction processes depend on the particular molecular structure of the components and their surfaces, as well as on local neighborhoods and energetic conditions.

The processes introduced above trigger the degradation of the lithium-ion battery. The degradation depends on several input parameters such as temperature [2], state of charge, relaxation time [3], utilization mode, and applied current [4, 5, 6]. The ageing originates from many different processes and their interactions [7]. In general, calendar and cycling ageing are distinguished. Both can lead to capacity or power fade of the battery due to loss of cyclable lithium, loss of active material and increased impedance. A brief overview of lithium-ion battery degradation is given in the following.

Many reported degradation processes are related to the SEI. Details are shown in chapter 3. Ageing in the context of the SEI is related to the ongoing growth of the film [4], the passivation of the electrochemically active surface [6], the change of

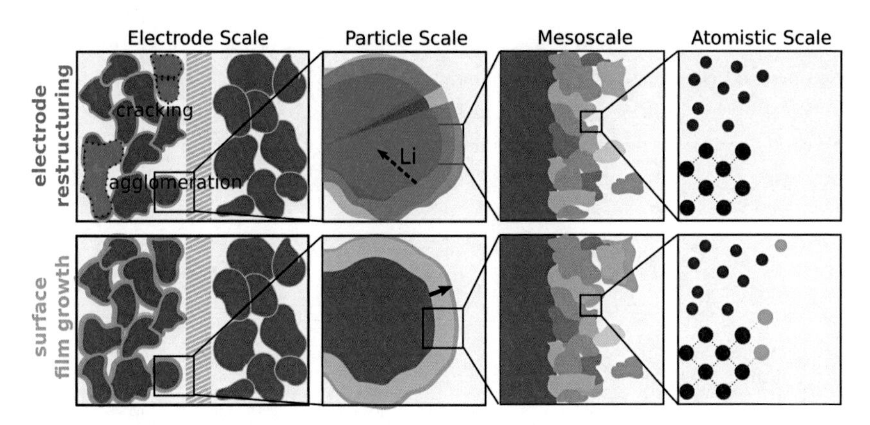

Figure 1.3: Illustration of the degradation of a lithium-ion battery through electrode restructuring (top) and surface film growth (bottom).

morphology and composition [7, 5, 4], the positive metal intercalation [7], and the cracking reconstruction cycle [7, 5, 4]. The surface film growth is illustrated in Figure 1.3 (bottom). Surface film growth is caused by deposition of the solid byproduct of the side reactions on the surface. On the electrode scale, the growth of the surface film will decrease the porosity of the electrode and thus impact diffusion and ion conduction. On the particle scale, film growth will decrease ion and electron conduction through the film and in between particles. On the mesoscale, heterogeneous film growth causes blocking or changing of the transport pathways through the film. On an atomistic scale, the surface film/electrolyte interface changes, which has an impact on the lithium-ion transfer between film and electrolyte and the side reactions on the surface film. An initial film is created during the very first cycles of the battery due to the dominance of the side reactions. The film formation is usually performed prior to use, thus being part of the production process. It is crucial for the properties of the film, and thus the performance and degradation of the battery. This production step is called the formation process.

Other degradation processes are related to the restructuring of electrodes, which can cause contact loss of active material, cracking of particles or agglomeration of particles [7, 8, 9, 10, 11]. Details are shown in chapter 5. Electrode restructuring is mainly triggered by the lithium storage in the active material host structures. This causes volume expansion of the material. In particularly high charging rates, electrodes and particles experience significant mechanical stress. The electrode restructuring is illustrated in Figure 1.3 (bottom). On the electrode scale, electrode restructuring leads to a change in lithium-ion diffusion and electron conduction and can cause the disconnection of particles. On the particle scale, the length of the diffusion pathways can increase or decrease through agglomeration and cracking of particles, respectively. Whereas surface film degradation usually takes place only at the negative electrode, electrode restructuring of the electrodes can be found at both the negative and the positive electrodes. As shown above, processes due to side reactions and operational

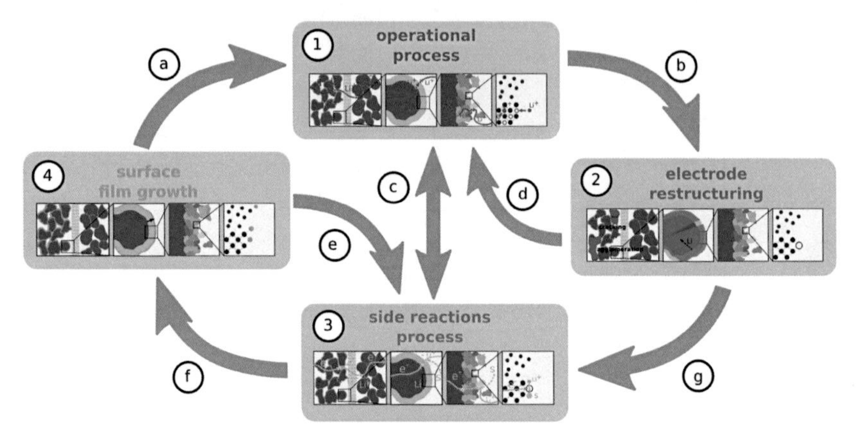

Figure 1.4: Illustration of the interaction between processes during operation and due to side reaction, with degradation through surface film growth and electrode restructuring. Descriptions of the interactions at positions a-g are provided in Table 1.1.

processes are both influenced by various aspects, from the electrode to the atomistic scale. Thereby, they trigger degradation processes, e.g. film growth or electrode restructuring. An illustration of the interaction between those aspects is provided in Figure 1.4. Descriptions of the interaction at positions a-g are summarized in Table 1.1. It can be seen that all aspects of the operation and degradation of lithium-ion batteries are strongly interconnected and often affect or are affected by various aspects across the scales. To summarize, this review clearly shows the need for sophisticated models describing those phenomena and their interactions. In the following section, the presently available models on the various scales involved are discussed.

1.3 Physical modeling of batteries [2]

Decades of research has aimed to develop models for lithium-ion batteries, which allow for more efficient computation of phenomena, more detailed physical insight,

[2]Part of this section has been published in (Krewer et al., J. Electrochem. Soc., 165:A3656–A3673, 2018 [12])

Table 1.1: Description of the interaction illustrated in Figure 1.4.

Pos.	Description
a	reduction of porosity, increase of ionic film resistance, change of transport pathways in SEI, modification of molecular structure of materials and their surface
b	volume expansion of electrodes, volume expansion of particles, stress on films, stress on molecular bonds
c	change of electrical potential and concentration, change of lithium concentration in the particles, change of composition within film components, depletion or flooding of species on film surface
d	disconnection of particles, change of electrochemically active surface areas, blocking of pores, change of diffusion pathways in particles, change of transport pathways in SEI
e	reduction of porosity, increase of electronic film resistance, change of transport pathways in SEI, modification of molecular structure of materials and their surfaces
f	production of solid species
g	disconnection of particles, change of electrochemically active surface areas, blocking of pores, change of transport pathways in SEI, exposition of uncovered surfaces, coverage of surfaces

better resolution of heterogeneity, and simulation of actual geometries. This section provides an overview of the most commonly used modeling approaches for lithium-ion batteries and thereby introduces the most important physical phenomena and available simulation methods. Physical phenomena can be observed at different time and length scales, as discussed in the previous section. This literature review is structured with respect to these length scales. A more quantitative definition for the scales is given here as

- Electrode scale: Electrochemical sandwich consisting of anode, cathode, and separator (10^{-6}-10^{-3} m)
- Particle scale: Primary structure of the active material (10^{-7}-10^{-5} m)
- Mesoscale: Structuring of material (10^{-9}-10^{-6} m)
- Atomistic scale: Atoms/Molecules (10^{-10}-10^{-7} m)

The transition between the single scales is smooth, of course, and thus the given groups are somewhat arbitrary. Nevertheless, it is reasonable in consideration of the actual structure of batteries. A slightly deviating definition can be found in literature [13, 14]. Different definitions of multiscale simulation are often used. In this work, every simulation which covers more than one of the aforementioned scales is considered as multiscale simulation. For simulation of processes on these scales, different modeling paradigms can be used. Frequent paradigms including definition and fundamental literature are:

- Continuum: Set of ODEs, PDEs, and algebraic equations (*Electrochemical Systems* [15] and *Porous-Electrode Theory with Battery Applications* [16])
- Monte Carlo: Methods based on random numbers, e.g. kinetic Monte Carlo (kMC) Method (*First-Principles Kinetic Monte Carlo Simulations for Heterogeneous Catalysis* [17] and *An Overview of Spatial Microscopic and Accelerated KMC Methods*[18])
- Molecular Dynamics: Simulations of atom and molecule interaction (*Molecular Dynamics* [19])
- Density Functional Theory: Determining system states based on quantum mechanics (*Density Functional Theory* [20])

Some of these methods are available in very different realizations and may be used in combination with even more fundamental methods, such as quantum mechanics. As soon as more than one method is used, this is considered to be multiparadigm modeling. In Figure 1.5 the accessible length scales for the different simulation paradigms are illustrated. In the following, the state of art for lithium-ion battery modeling in general and efforts in muitscale modeling in particular is given.

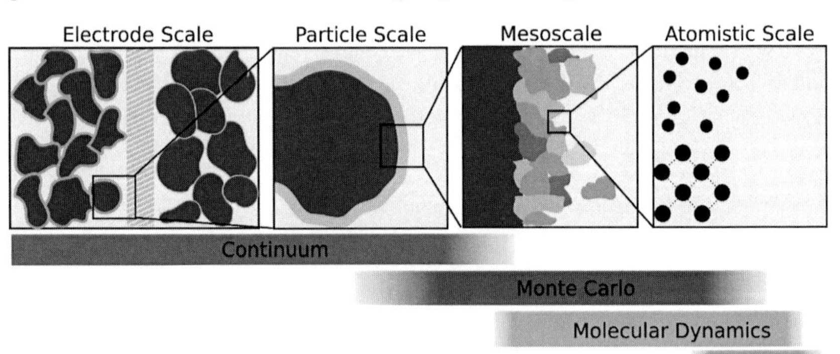

Figure 1.5: Scales of the various modeling paradigms.

The most commonly used physical model is the pseudo 2-dimensional model (P2D) as given in [21, 22], first introduced in 1993 by Doyle et al. [21]. The general structure of the P2D model is illustrated in Figure 1.6. Together, good agreement to experimental data and computational efficiency allow for simulations from short to long time scales, making a good compromise for a wide range of applications and cell types. The model includes the lithium-ion diffusion and electrical charge transport along electrode thickness in one dimension, x, and the diffusion of lithium into the active material particles in a second dimension, r, as well as the electrochemical charge transfer reactions at both electrodes. During the insertion of lithium into the solid active material, the electrochemical potential changes, which is often considered to be an empirical function of the equilibrium potential. Electrodes are thereby modeled as homogeneous multiphase systems based on the porous electrode theory [16]. With this setup the P2D model covers two scales, i.e. the electrode scale and the particle scale. Models with similar structure but particular modification are widely used and under development. In the following, some common modifications, including chemical and mechanical degradation, as well as heat transport and dynamic operation, are discussed.

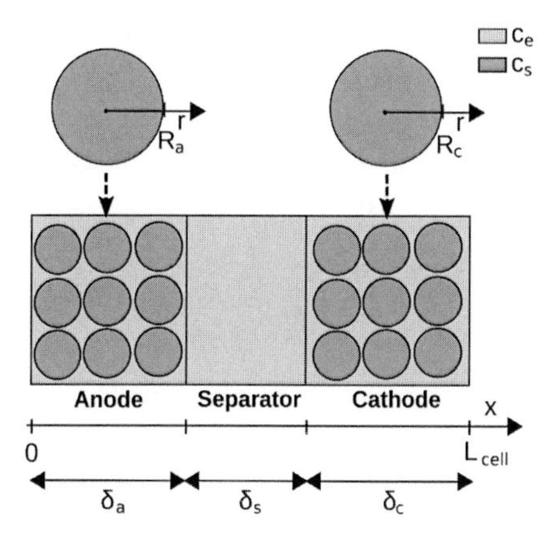

Figure 1.6: Illustration of the P2D model.

Inclusion of fast dynamic processes at the electrochemical double layer [23, 24] facilitates the simulation of electrochemical impedance spectroscopy [25, 26, 27]. The most frequently investigated degradation process is the formation of the SEI due to operation outside the electrochemical stability window of the electrolyte at the negative electrode. This process denotes loss of cyclable lithium, which mainly leads to a depletion of lithium in the positive electrode and the unbalancing of the system. The reaction products form a layer at the surface of the negative electrode particles. This process is included with varying amounts of detail and under different assumptions in several models of the P2D type [28, 29, 30, 31, 32, 33, 34]. Due to volume change of active material during lithiation/delithiation, particles and electrode structures are stressed mechanically, which is addressed in various models [35, 36]. Volume expansion can cause the destruction and reformation of the SEI [37, 38]. In general, the P2D model can be used for parameter identification based on electrochemical measurements, such as charge/discharge curves [39] or electrochemical impedance spectroscopy [40].

Several approaches for the reduction of the P2D model are possible. Model reduction is motivated by the need for computationally faster models. A widely applied reduction is the approximation of the whole insertion process of the electrode by a single particle and leads to the single particle model [41, 42, 43]. This is a good approximation as long as the electrode states, e.g. electrolyte concentration or electrical potential, are homogeneous along the thickness of an electrode. This is often the case for slow charging/discharging, thin film electrodes [44] or fast ionic and electrical conductivity. Another model reduction is the approximation of the diffusion process into the active material. Approximation can be a polynomial [41], an analytic solution or neglecting the concentration gradient with a zero dimensional model, which can be sufficiently accurate for active materials with fast diffusion or considerably small particles. In general, thermal aspects [45], chemical [46] and mechanical degradation, and double layer capacitance can be considered. With simplifications as shown here, computational efficiency of the model can be increased considerably.

Actual technical batteries are usually composed of several cells in parallel or row. The particular three dimensional compartment can considerably impact the performance of the system. Transport of electrical charge and heat conduction are the major processes on this length scale. Efforts have been made to simulate the effect of battery stacking [47] or the effect of temperature distribution in a 3-dimensional cell by coupling thermal and electrochemical equations[48, 49, 50, 51]. Higher dimensional models are computationally much more expensive and thus cannot be used for online prediction or state estimation. Rather, they are applied to understand the impact of cell design and temperature distribution on the cell performance.

Porous electrodes are highly heterogeneous systems, e.g showing heterogeneous pores and particle sizes, a heterogeneous composition and arrangement in the electrode. In the P2D model, this heterogeneity is homogenized based on porous electrode theory. Even though this homogenization is often sufficient for reproducing the overall performance of the electrode, it is often insufficient for diagnosing and predicting electrode performance for specific structures and for understanding certain degradation effects triggered by local heterogeneity, e.g. local current densities or temperatures. As a consequence, much effort has been made to include heterogeneity in electrode models. In general, explicit and statistical heterogenization can be distinguished. There are models available which explicitly model the transport and reaction phenomena in consideration of a distinct microstructure in 2D [52, 53] and 3D [54, 55, 56]. Since the resolution of the whole electrode is computationally very expensive in some models, only a representative part of the microstructure is simulated and coupled either directly [57] or indirectly, through a surrogate model [58], with a homogeneous model. Further, models are available which consider details about heterogeneous particle-particle interaction and electron transport [59]. To avoid explicit consideration of complex structures, efforts have been made to consider particle size distribution as statistical property in homogeneous electrochemical models [60, 61, 40, 62, 63]. In general, models including heterogeneity are computationally more expensive compared to homogeneous models, but they enable a more detailed insight for diagnosis. Statistical heterogenization or surrogate models can nevertheless enable to include this information with a reasonable computational effort.

Some key processes in batteries are on much smaller length scales, e.g. on mesoscale, and thus can barely be modeled by continuum approaches, although they determine the overall macroscopic processes. There is a need to understand the impact of these processes and to enable knowledge based design and control to avoid degradation and failure, and increase energy or power density. One important dynamic process that can be considered as a mesocale process is the charging of the electrochemical double layer. It is usually considered on the electrode scale, based on simple electrical analogy models. On the mesoscale, work has been done to include the atomistic effects of the double layer in non-equilibrium conditions [64] or its 3 dimensional modeling [65]. The kMC method has shown to be an adequate technique to resolve processes on an atomistic scale, while still enabling longer simulation horizons compared to other atomistic techniques. Further, kMC enables the investigation of structures on the mesoscale, such as surface films or nano pore systems. Blanquer et al. [66] applied kMC to lithium air batteries and simulated O_2 and lithium transport inside a pore. They used this model because the continuum model is not able to capture important effects in those small pores. KMC has been applied to simulate diffusion of lithium and electrons through different active material host structures [67, 68, 69, 70]. Further, in [71] the experimentally observed morphology of Si nanowires can be explained by kMC simulation. In the work shown by Methekar et al. [72], it is applied to investigate passivation of the negative electrode surface through side reactions. A similar approach is shown in [73] for simulation of an initial growth of the SEI. Kim et al. [74] used Molecular Dynamics (MD) to study electrolyte degradation on a Li or graphite surface, which enables one to resolve the mesostructre of the SEI. Nevertheless, MD is usually not sufficiently fast to compute the dynamics of mesostructures. KMC facilitates such simulation, but it is not frequently used in lithium-ion batteries. This is despite the fact that mesoscale problems are important for battery operation and degradation.

MD and DFT are widely used simulation paradigms for the simulation of atomistic problems. In particular, for chemical degradation in context of the SEI, MD has been applied frequently to determine the respective reaction pathways[75, 76, 77, 78]. Further, this technique has been used to study the diffusion dynamic of lithium [79]. DFT has been applied to determine the activation energies for reaction pathways [80] or to calculate the open circuit potentials of active materials [70, 81]. Further, it was used to study diffusion dynamics of lithium or electrons through different SEI components [82, 83, 84, 85]. Whereas atomistic methods are widely used for detailed understanding and diagnosis of particular atomistic processes, there is a need to transfer such knowledge from atomistic scale methods to larger scale methods.

As discussed previously, most phenomena of lithium-ion batteries cannot be reduced to a single scale model. Therefore, there is a need for multiscale models. The aforementioned, and most frequently used, P2D model can be seen as a multiscale model between the electrode and the particle scale. Those models were included in cell models to bridge the scale to the cell level or were heterogenized to more accurately include the heterogeneous nature between the scales. When the difference in scale is too large, the application of different simulation paradigms is needed. As a consequence, there is a need for new coupling strategies to enable direct multiparadigm simulation. One example is the coupling of continuum and kMC models, as it has been applied to copper deposition [86], fuel cells [87], and others [88] already.

1.4 Scope of this work

As shown above, research has been carried out in modeling and simulation on different scales using different paradigms. Thereby, degradation problems are investigated on all scales. Nevertheless, the actual multiscale nature as examined in section 1.2 is barely addressed. To investigate multiscale interactions, models should cover macroscopic transport processes in heterogeneous electrodes, while considering the effect of mesostructures and heterogeneous reactions at the surface, in order to link macroscopic and atomistic processes. Adequate models are presently not available, which impedes the transfer of knowledge from fundamental physics and the study of multiscale interactions.

The objective of this thesis is the development of methods for multiscale modeling of degradation problems for lithium-ion batteries as introduced in section 1.2. To cover those degradation processes and their impact on the multiple scales, continuum models are combined with kMC modeling approaches, which allows for the consideration of processes from electrode down to meso- and atomistic scale. Special focus is laid on the development of adequate models and algorithms, which are presently not available. This work thereby covers all aspects shown in Figures 1.2 and 1.3 and most of the interactions illustrated in Figure 1.4. Examples are mostly shown for issues related to the SEI on negative electrodes, because of their importance and the good situation concerning experimental and theoretical work in literature. The thesis contains the following parts:

- Direct Multiparadigm Algorithms (chapter 2)
- Multiparadigm Simulation of Heterogeneous Film Growth Mechanisms (chapter 3)
- Multiscale Analysis of Film Formation (chapter 4)
- Macroscopic Heterogeneity (chapter 5)

There are various options for the design algorithms for coupling continuum and kMC models. To allow for a directed selection and tuning of the algorithm, in chapter 2 different algorithms are compared and evaluated concerning computational cost and accuracy. For this, a simple example problem including electrochemical reactions on an active surface is implemented. By excluding higher order reactions and interaction on the surface, the simulation can be solved accurately with a pure continuum code, which serves as a benchmark for accuracy of the multiscale solution. The results show the basic differences of the investigated algorithms and the general effects of certain tuning aspects, e.g. smoothing and field size.

In chapter 3, the multiparadigm technique is applied to surface film formation in a lithium-ion battery. A single particle model of an electrode is thereby extended by a SEI growth model using the kMC method. An exemplary reaction mechanism is implemented that considers decomposition of an ethylene carbonate based electrolyte at a graphite electrode. Simulations are carried out for two electrodes with different particle sizes. It is demonstrated how this technique can be applied to an actual battery operation. The presented model enables the simulation of the typical features of the potential slope during the first charging of batteries, the so called formation process. It allows one to analyze the evolution of thickness and structure of the surface film during operation. With this, the chapter covers the interactions at positions a,c,e, and f, as illustrated in Figure 1.4.

While the technique was established in chapter 3 for the simulation of surface film growth problems, chapter 4 goes one step further, enabling a multiscale analysis of surface film formation at actual technical cells. Therefore, the concept is implemented in the state of the art battery model, which is the P2D model developed by Doyle and Newman. Further, a concept for parameter identification is presented, while the need of a thermodynamically consistent model formulation is emphasized. Basic steps are presented using experimental data. The capability for multiscale analysis is demonstrated with this parameterized model and distinct multiscale effects are revealed for the analyzed cell. This motivates the application of such analysis in future research.

In previous chapters electrodes were treated as homogeneous systems even though they are highly heterogeneous; it can be expected that this heterogeneity may trigger degradation processes such as surface film formation. The effect of macroscopic heterogeneity is investigated by the example of particle size distribution. The results of chapters 3 and 4 indicated that the local current densities and system states impact the local surface current densities and film growth. Therefore, in chapter 5 the effects of different particle size distributions on the performance and degradation are investigated. For this, the homogeneous single particle model is extended to consider heterogeneity as particle size distribution. Further, a change in the particle size distribution is investigated by applying population balance equations. Results show that the local current densities depend on the particle size distribution and that the change of macroscopic heterogeneity impacts the c-rate capability of the electrode. With this, this chapter covers the remaining interactions at positions b, c, and g, as illustrated in Figure 1.4.

Chapter 2

Direct Multiparadigm Algorithms [3]

2.1 Introduction

In this chapter, the concept and algorithms for multiparadigm simulation is introduced. Further numerical aspects, e.g. accuracy and computational cost, are discussed. The methodology is applied in chapters 3 and 4 for simulation and analysis of multiscale phenomena in batteries.

The demand for computational efficiency often leads to the application of simplified physical or phenomenological homogenized continuum models. However, for many systems, in particular those containing active surfaces, their behavior is determined by complex mechanisms that occur at the atomistic scale, such as electrochemical surface degradation in batteries or fuel cells. In order to facilitate a knowledge-driven optimization of such systems, atomistic effects need to be included in macroscopic models, which leads to an increasing need for multiscale simulations [14]. Simulation techniques applicable for small length scales, such as the kinetic Monte Carlo (kMC) method, are often of a stochastic nature and cannot be simulated for the larger length and time scales describable by continuum models. Therefore, multiparadigm approaches are used to bridge those scales. The coupling of scales can be realized directly or indirectly [13]. Only direct coupling allows for the investigation of the interaction of the models and phenomena between the scales. Direct coupling of different simulation paradigms is challenging due to the fundamentally different representation of physical processes at different scales. Coupling algorithms need to be further developed in order to improve accuracy and computational efficiency [90]. Better algorithms will help to further spread the application of those techniques, which promise to lift electrochemical modeling to a more advanced and accurate stage for the description of physical processes. The narrowing of the gap between engineering and computational chemistry is important for electrochemical systems in particular,

[3] Part of this chapter has been published in (Röder et al., Comput. Chem. Eng., 121, 722–735, 2019 [89])

and heterogeneous catalysis in general [91].

Here, methods for the direct coupling of kMC and continuum models based on differential equations are investigated. KMC has been shown to be a useful tool to investigate surface chemistry for many years [88]. An excellent overview on concepts, status, and challenges of applying first-principles kMC to reactions at surfaces is given in [17]. In electrochemistry, for instance, kMC has been applied to investigate CO electrooxidation [92], copper electrodeposition [93], fuel cells [94], and batteries [66, 72]. In contrast to continuum models, kMC allows the study of lateral stochastic interactions of molecules on surfaces at the atomistic scale [95].

KMC models can be coupled with continuum models in a multiparadigm model [95, 14, 13]. The importance, perspectives, and challenges of such coupled models have been outlined by various researchers, e.g., [95, 90, 13, 14, 96]. Multiparadigm algorithms have been applied to study the agglomeration of particles in reactors by coupling computational fluid dynamics (CFD) with Monte Carlo methods [97], CO oxidation in a catalytic reactor by coupling CFD and first principles kMC [98], in the additive-mediated electrodeposition of copper [86], and to fuel cells and other energy storage systems [99, 87]. In the field of batteries, mulitparadigm approaches have been used to study the formation of the solid electrolyte interface on negative electrodes in lithium-ion batteries [72, 100].

In general, direct coupling – also called hybrid or heterogeneous-homogeneous coupling – and indirect coupling can be distinguished. Direct coupling includes a frequent interaction of continuum model and kMC model during simulation [13]. Designing a numerically stable and accurate direct multiparadigm algorithm is challenging, and so an indirect coupling strategy is preferred if such a formulation can be derived that sufficiently captures the underlying coupled phenomena. Some multiscale systems, however, need direct coupling to accurately describe their behavior [14]. Direct coupling separates the simulation into time intervals where the kMC and the continuum model are solved in sequence or in parallel [96]. Both algorithms can be iterated multiple times within each time interval convergence to self-consistent solutions between the models [101]. Further, filtering techniques are often applied to reduce fluctuations of the kMC output [93, 102]. All aspects have considerable impact on accuracy and computational cost of the simulation, which is rarely addressed. Theoretical approaches are available [103, 104, 105], but yet no comprehensive comparison is available between different strategies and their impact on accuracy and computational cost.

In this chapter, different algorithms for direct coupling of kMC with continuum models provided and evaluated. The focus is on electrochemical interfaces, which arise in many systems of scientific and engineering interest. To benchmark the performance of the algorithms, a simple example problem is defined with sequential reactions with only first-order reaction kinetics including an electrochemical reaction step. By excluding heterogeneous processes, i.e. atom to atom interaction, the problem can be accurately solved using mean field approximation in pure continuum codes, which enables the evaluation of the accuracy of the multiparadigm simulations. The aim is to provide an introduction to direct multiparadigm algorithms applicable to surface degradation problems in electrochemical engineering. Further, the impact of grid size, time step length, and smoothing of fluctuations on the performance, i.e. computational cost and accuracy, of the algorithms is shown.

2.2 Computational details

2.2.1 Example problem

For the purpose of evaluating multiscale modeling algorithms, a simple example problem is introduced, which includes a Fmechanism of two consecutive reactions of first order, without interaction of the species on the surface. The mechanism is illustrated in Figure 2.1 (A) and the considered processes are given in Table 2.1. Process I is adsorption of species A^+ from the liquid phase, i.e. the electrolyte, to the surface. Process II is the electrochemical reaction of species A^+ to species B. Process III is the subsequent chemical reaction of species B to species C. Finally, Process IV is the desorption of species C from the surface. All processes are reversible. The presented example problem does not include any atom to atom interaction and can thus also be solved exactly by a pure continuum code, allowing the validation of the multiparadigm code and benchmark of the different multiparadigm algorithms.

With the here presented multiparadigm approach, a continuum model composed of ordinary differential equations, and an atomistic model using the kMC method, are directly coupled. The thermodynamics and kinetics of each process in each model are detailed below. It is suggested, that only the processes of particular interest or clearly heterogeneous nature should be modeled on an atomistic scale (e.g., [86]). All other processes, which can be sufficiently accurately approximated by mean field approaches, should preferably be solved by the much faster continuum codes. The boundary between models is defined between educts and products of a process, which is illustrated in Figure 2.2 for process II. Species A^+ and e^- are considered by the continuum model and species B is considered by the kMC model. The models are coupled by synchronization of the fluxes, e.g. the flux of process II. In general, the boundary between continuum and kMC model can be flexibly defined at every considered process. The choice will depend on the investigated problem.

In the following, first the continuum and the kMC model is provided. Then, the sequence concept and the multiparadigm algorithms for coupling of those models are introduced. Finally, the approach to quantify simulation errors and the investigated simulation scenario are given.

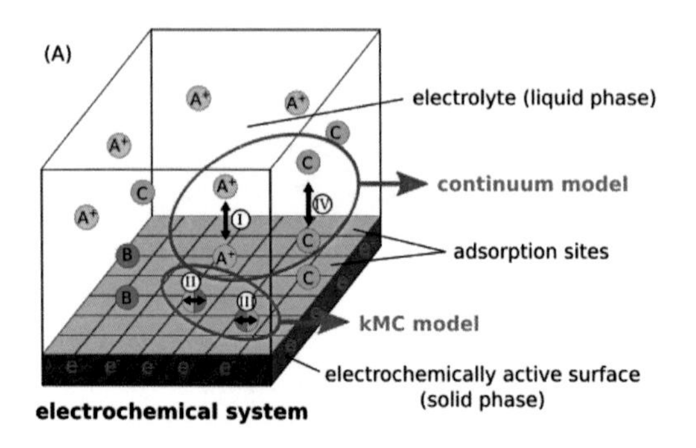

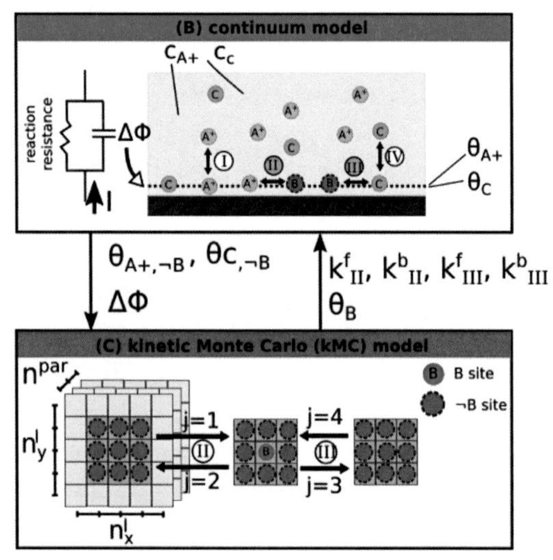

Figure 2.1: Example of the electrochemical system (A) and illustration of the multiparadigm model including the continuum model (B) and the kMC model (C). Species and processes not addressed in a model are colored in grey. Reprinted from publication [89] with permission from Elsevier.

2.2.2 Continuum model

The scope of the continuum model is illustrated in Figure 2.1 (B). It features changes of electrical potential, $\Delta\Phi$, at the interface, changes of concentrations of species A^+

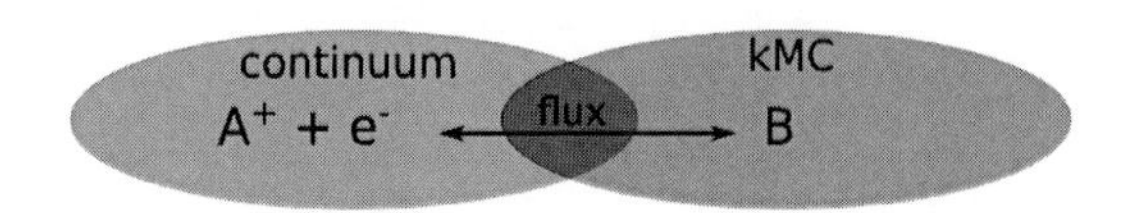

Figure 2.2: Illustration of the interface between kMC and continuum models. Reprinted from publication [89] with permission from Elsevier.

Table 2.1: Reactions and adsorption/desorption processes I-IV, with continuum processes and their rate constants k and the processes j as considered in the kMC model with their microscopic rates Γ. Reprinted from publication [89] with permission from Elsevier.

Nb.	Continuum Process	j	kMC Process
I	$A^+(E) + V(ads) \underset{k_I^b}{\overset{k_I^f}{\rightleftharpoons}} A^+(ads)$	–	–
		–	–
II	$A^+(ads) + e^-(el) \underset{k_{II}^b}{\overset{k_{II}^f}{\rightleftharpoons}} B(ads)$	1	$A^+(ads) + e^-(el) \xrightarrow{\Gamma_i^{l,1}} B(ads)$
		2	$B(ads) \xrightarrow{\Gamma_i^{l,2}} A^+(ads) + e^-(el)$
III	$B(ads) \underset{k_{III}^b}{\overset{k_{III}^f}{\rightleftharpoons}} C(ads)$	3	$B(ads) \xrightarrow{\Gamma_i^{l,3}} C(ads)$
		4	$C(ads) \xrightarrow{\Gamma_i^{l,4}} B(ads)$
IV	$C(E) + V(ads) \underset{k_{IV}^b}{\overset{k_{IV}^f}{\rightleftharpoons}} C(ads)$	–	–
		–	–

and C, $c_{A^+(E)}$ and $c_{C(E)}$, within the electrolyte, and changes of surface fractions of species A^+ and C, $\theta_{A^+(ads)}$ and $\theta_{C(ads)}$, on the adsorption site. Surface fraction of B, $\theta_{B(ads)}$, is an input parameter provided by the kMC model. Further, the model covers expressions for processes I-IV, as illustrated in Figure 2.1 (B). While processes I and IV are independent of the kMC model, processes II and III are synchronized with the kMC model by adapting the forward and backward reaction rate constants of reactions II and III, namely k_{II}^f, k_{II}^b, k_{III}^f, k_{III}^b. It is noted that instead of passing reaction rate constants, reaction fluxes can also be passed from the kMC to the continuum model. However, passing rate constants yields more numerically stable codes, as discussed below. In the following, the equations of the continuum model are given.

The charge balance is included as

$$C^{DL}\frac{d\Delta\Phi}{dt} = I - q_{II}F \tag{2.1}$$

with applied electrical current I, surface flux of the electrochemical reaction II, q_{II}, double layer capacitance C^{DL} and Faraday constant F. Balance equations for surface fraction of species A^+ and C are given as

$$\frac{N_s}{o_s}\frac{\mathrm{d}\theta_{\mathrm{A^+(ads)}}}{\mathrm{d}t} = q_{\mathrm{I}} - q_{\mathrm{II}} \tag{2.2}$$

and

$$\frac{N_s}{o_s}\frac{\mathrm{d}\theta_{\mathrm{C(ads)}}}{\mathrm{d}t} = q_{\mathrm{III}} + q_{\mathrm{IV}}, \tag{2.3}$$

respectively, using site density N_s, site-occupancy number o_s and fluxes q of processes I-IV. Balance equations for concentrations of A^+ and C within the electrolyte are included as

$$\frac{1}{a_s}\frac{\mathrm{d}c_{\mathrm{A^+(E)}}}{\mathrm{d}t} = q_{\mathrm{I}} \tag{2.4}$$

and

$$\frac{1}{a_s}\frac{\mathrm{d}c_{\mathrm{C(E)}}}{\mathrm{d}t} = q_{\mathrm{IV}}, \tag{2.5}$$

respectively, using specific surface area a_s. Fluxes q of processes I-IV are provided by

$$\frac{o_s}{N_s}q_{\mathrm{I}} = k_{\mathrm{I}}^f\theta_{\mathrm{V}}\frac{c_{\mathrm{A^+(E)}}}{C^0_{\mathrm{A^+(E)}}} - k_{\mathrm{I}}^b\theta_{\mathrm{A^+(ads)}}, \tag{2.6}$$

$$\frac{o_s}{N_s}q_{\mathrm{II}} = k_{\mathrm{II}}^f\theta_{\mathrm{A^+(ads)}}\exp\left(\frac{\beta\Delta\Phi\mathrm{F}}{\mathrm{R}T}\right) - k_{\mathrm{II}}^b\theta_{\mathrm{B(ads)}}\exp\left(-\frac{(1-\beta)\Delta\Phi\mathrm{F}}{\mathrm{R}T}\right), \tag{2.7}$$

$$\frac{o_s}{N_s}q_{\mathrm{III}} = k_{\mathrm{III}}^f\theta_{\mathrm{B(ads)}} - k_{\mathrm{III}}^b\theta_{\mathrm{C(ads)}}, \tag{2.8}$$

and

$$\frac{o_s}{N_s}q_{\mathrm{IV}} = k_{\mathrm{IV}}^f\theta_{\mathrm{V}}\frac{c_{\mathrm{C(E)}}}{C^0_{\mathrm{C(E)}}} - k_{\mathrm{IV}}^b\theta_{\mathrm{C(ads)}}, \tag{2.9}$$

with forward and backward reaction rate constants, k^f and k^b, symmetry factor, β, standard state concentrations, C^0, and temperature, T. Reaction II is an electrochemical reaction and thus includes an exponential dependency of electrical potential $\Delta\Phi$ at the interface. Surface fraction θ_{B} and reaction rate constants k_{II}^f, k_{II}^b, k_{III}^f and k_{III}^b are provided as input parameters from the kMC model as given in the following sections. Surface fractions of vacant sites V(ads) are determined as

$$\theta_{\mathrm{V}} = 1 - \theta_{\mathrm{A^+(ads)}} - \theta_{\mathrm{B(ads)}} - \theta_{\mathrm{C(ads)}}. \tag{2.10}$$

Forward and backward rate constants of the sorption processes I and IV are independent of the kMC model and determined by the following equations:

$$k_{\mathrm{I}}^{f} = k_{\mathrm{I}} \exp\left(\frac{-E_{\mathrm{I}}^{A}}{\mathrm{R}T}\right), \tag{2.11}$$

$$k_{\mathrm{I}}^{b} = k_{\mathrm{I}} \exp\left(\frac{-(E_{\mathrm{I}}^{A} - \Delta G_{\mathrm{I}}^{0})}{\mathrm{R}T}\right), \tag{2.12}$$

$$k_{\mathrm{IV}}^{f} = k_{\mathrm{IV}} \exp\left(\frac{-E_{\mathrm{IV}}^{A}}{\mathrm{R}T}\right), \tag{2.13}$$

and

$$k_{\mathrm{IV}}^{b} = k_{\mathrm{IV}} \exp\left(\frac{-(E_{\mathrm{IV}}^{A} - \Delta G_{\mathrm{IV}}^{0})}{\mathrm{R}T}\right) \tag{2.14}$$

with activation energy of the forward process, E^{A}, and standard state Gibbs free energy of the forward process, ΔG^{0}.

2.2.3 Kinetic Monte Carlo model

The scope of the kMC model is illustrated in Figure 2.1 (C). The model includes the reaction processes II and III. Only species B is considered explicitly, i.e. sites covered with B. The species A^+ and C are assumed to be homogeneously distributed on the sites that are not covered by B, referred to as $\neg$B site. The fraction of surface species A^+(ads) on $\neg$B sites, $\theta_{\mathrm{A^+(ads)},\neg\mathrm{B}}$, surface species C(ads) on $\neg$B sites, $\theta_{\mathrm{C(ads)},\neg\mathrm{B}}$, and the electrical potential, $\Delta\Phi$, are provided as input parameters from the continuum model. Details about model interaction are provided in the next section.

Kinetic equations for reaction II and III are solved using a rejection-free kMC algorithm with variable step size. Steps within the algorithm are explained in the following. In general the kMC algorithm includes the following actions with very kMC step i:

1. Calculate microscopic rate $\Gamma_i^{l,j}$ of all microscopic processes $j \in \{1,2,3,4\}$ on every lattice sites $l \in \{z | z \in \mathbb{N}, z \leq n^l\}$
2. Calculate the step time length Δt_{i+1} applying a random number $\zeta_1 \in (0,1)$
3. Select one microscopic process $J_i \in \{1,2,3,4\}$ and one lattice site $L_i \in \{z | z \in \mathbb{N}, z \leq n^l\}$ taking into account the microscopic rates $\Gamma_i^{l,j}$ applying a second random number $\zeta_2 \in (0,1)$
4. Perform the selected process J_i on surface site L_i

5. If $i < \iota$, with ι being the last step of the kMC sequence, go to 1

Thereby, the kMC model sets up a cubic lattice with $n^l = n^l_x \cdot n^l_y$ lattice sites, with lattice state $\vartheta^l_i \in \{0, 1\}$ at lattice site l. A lattice site covered by B yields $\vartheta^l_i = 1$, i.e. B site, and a lattice site not covered by B yields $\vartheta^l_i = 0$, i.e. $\neg$B site. Within the kMC algorithm, in every kMC step i lattice state is changed. The lattice site is selected according to microscopic rates of the considered microscopic processes. A lattice site transfers $\vartheta^l_i = 0$ to $\vartheta^l_i = 1$ with the microscopic rate

$$\Gamma^{l,1}_i = (1 - \vartheta^l_i) k_{\mathrm{II}} \theta_{\mathrm{A^+(ads)},\neg\mathrm{B}} \exp\left(\frac{-E^A_{\mathrm{II}}}{RT}\right) \exp\left(\frac{\beta \Delta\Phi \mathrm{F}}{RT}\right) \tag{2.15}$$

corresponding to a microscopic forward step of process II and

$$\Gamma^{l,4}_i = (1 - \vartheta^l_i) k_{\mathrm{III}} \theta_{\mathrm{C(ads)},\neg\mathrm{B}} \exp\left(\frac{-(E^A_{\mathrm{III}} - \Delta G^0_{\mathrm{III}})}{RT}\right) \tag{2.16}$$

corresponding to a microscopic backward step of process III. Accordingly, a lattice site transfers from $\vartheta^l_i = 1$ to $\vartheta^l_i = 0$ with the microscopic rate

$$\Gamma^{l,2}_i = \vartheta^l_i k_{\mathrm{II}} \exp\left(\frac{-(E^A_{\mathrm{II}} - \Delta G^0_{\mathrm{II}})}{RT}\right) \exp\left(\frac{-(1-\beta)\Delta\Phi \mathrm{F}}{RT}\right) \tag{2.17}$$

corresponding to microscopic backward step of process II and

$$\Gamma^{l,3}_i = \vartheta^l_i k_{\mathrm{III}} \exp\left(\frac{-E^A_{\mathrm{III}}}{RT}\right) \tag{2.18}$$

corresponding to a microscopic forward step of process III. With this, a total of $n^j = 4$ microscopic processes are considered in the kMC model. The microscopic rates include the following input values: $\Delta\Phi_s$, $\theta_{\mathrm{A^+(ads)},\neg\mathrm{B}}$ and $\theta_{\mathrm{C(ads)},\neg\mathrm{B}}$, which are constant within a kMC sequence. Updating those input values with every kMC time step i is computationally very expensive and thus not feasible. Assuming constant values causes an error in the coupled simulation, which is evaluated and discussed to assess the coupling quality as detailed below.

In every kMC step i, first the step time length is calculated based on a uniform distribution random number $\zeta_1 \in (0, 1)$ as

$$\Delta t^{\mathrm{kMC}}_{i+1} = \frac{-\ln(\zeta_1)}{\Gamma^{\mathrm{tot}}_i}, \tag{2.19}$$

with Γ_i^{tot} being the total microscopic rate, which is calculated as

$$\Gamma_i^{\text{tot}} = \sum_{l=1}^{n^l} \sum_{j=1}^{n^j} \Gamma_i^{l,j}. \tag{2.20}$$

The discrete time at the following kMC time step $i+1$ can be calculated with

$$t_{i+1}^{\text{kMC}} = t_i^{\text{kMC}} + \Delta t_{i+1}^{\text{kMC}}. \tag{2.21}$$

Further, in every kMC step, one of the n^j microscopic processes j and one of the n^l lattice sites l, is selected, with respect to the microscopic rate $\Gamma_i^{l,j}$ as given in Equations 2.15-2.18, applying a second uniform distribution random number $\zeta_2 \in (0,1)$ according to

$$\frac{\sum_{l=1}^{L_i} \sum_{j=1}^{J_i-1} \Gamma_i^{l,j}}{\Gamma_i^{\text{tot}}} < \zeta_2 \leq \frac{\sum_{l=1}^{L_i} \sum_{j=1}^{J_i} \Gamma_i^{l,j}}{\Gamma_i^{\text{tot}}} \tag{2.22}$$

with J_i being the selected process and L_i being the selected grid point. A boolean Ψ, which indicates the selection of a process j within a kMC step i is determined as

$$\Psi_i^j = \begin{cases} 1, & \text{if } J_i = j \\ 0, & \text{otherwise} \end{cases} \tag{2.23}$$

to enable counting how often a certain process j occurs.

2.2.4 Sequence and model interaction

As shown in the previous sections, both models rely on input of the respective other model. This requires to pass input parameters between the models, which needs to be handled by the MPA. One possibility is to split the simulation in time sequences. In each time sequence both models are simulated one after another with fixed sequence specific input parameters. Input of the model simulated first needs to be estimated, while the input of the second model can be evaluated using the results of the first model.

The simulation procedure during a sequence is illustrated in Figure 2.3. It can be seen that within a sequence s the continuum and the kMC model are solved. The sequence period is determined as

$$\Delta t_s^{\text{seq}} = \frac{t^{\text{end}}}{n^{\text{seq}}} \tag{2.24}$$

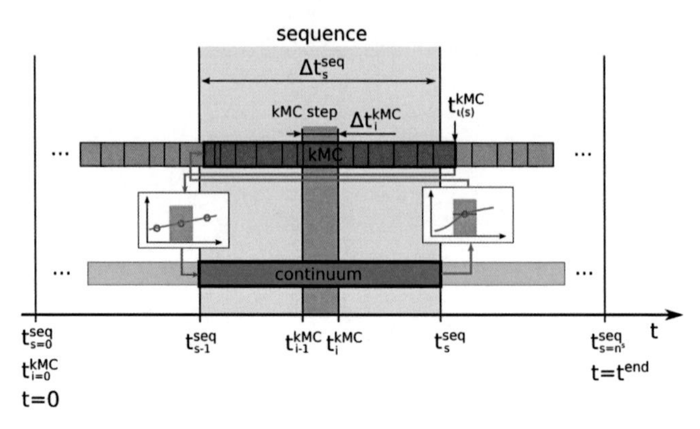

Figure 2.3: Processes during one sequence in the coupled simulation. Reprinted from publication [89] with permission from Elsevier.

where t^{end} is the simulation time and n^{seq} the number of sequences. This determines the end time of the following sequence $s+1$

$$t_{s+1}^{\mathrm{seq}} = t_s^{\mathrm{seq}} + \Delta t_{s+1}^{\mathrm{seq}} \tag{2.25}$$

The mean time corresponding to a sequence is defined as

$$\bar{t}_s^{\mathrm{seq}} = \frac{t_s^{\mathrm{seq}} + t_{s-1}^{\mathrm{seq}}}{2} \tag{2.26}$$

The time of the last kMC step i corresponding to a sequence s, ι, is not equal to but slightly higher than the end time of the sequence t_s^{seq}, as kMC step time length is determined using a random number. The last kMC step ι can be determined as a function of s, i.e. $\iota = f(s)$, according to

$$\iota = \max(i) \in \{i | t_{s-1}^{\mathrm{seq}} < t_{i-1}^{\mathrm{kMC}} \leq t_s^{\mathrm{seq}}\} \tag{2.27}$$

Further, the sequence ς, corresponding to of a certain point in time can be determined as a function of i, i.e. $\varsigma = f(i)$, according to

$$t_{\varsigma-1}^{\mathrm{seq}} < t_{i-1}^{\mathrm{kMC}} \leq t_{\varsigma}^{\mathrm{seq}} \tag{2.28}$$

The input parameters for the kMC model in step i are evaluated with the continuum model at the mean time $\bar{t}_\varsigma^{\text{seq}}$ of the sequence ς as

$$\theta_{\mathrm{A^+(ads)},\neg\mathrm{B}}(i) = \frac{\theta_{\mathrm{A^+(ads)}}(\bar{t}_\varsigma^{\text{seq}})}{1-\theta_{\mathrm{B(ads)}}(\bar{t}_\varsigma^{\text{seq}})} \tag{2.29}$$

$$\theta_{\mathrm{C(ads)},\neg\mathrm{B},s}(i) = \frac{\theta_{\mathrm{C(ads)}}(\bar{t}_\varsigma^{\text{seq}})}{1-\theta_{\mathrm{B(ads)}}(\bar{t}_\varsigma^{\text{seq}})} \tag{2.30}$$

$$\Delta\Phi(i) = \Delta\Phi(\bar{t}_\varsigma^{\text{seq}}) \tag{2.31}$$

which are functions of kMC time step i and constant within a sequence.

The input to the continuum model needs to be provided as a continuous function of time t. In the following, the required evaluation of kMC simulations to determine these inputs is shown. The kMC simulations are performed as parallel instances v, i.e. several kMC simulations with equal simulation input are performed in parallel during each sequence. This allows one to statistically evaluate the kMC output as shown below. The mean surface fraction of species B within one simulation instance v and one sequence s is determined as

$$\hat{\theta}_{\mathrm{B}}(s) = \frac{\sum_{i=\iota(s-1)}^{\iota(s)} \left(\Delta t_i^{\text{KMC}} \sum_{l=1}^{n^l} \vartheta_i^l\right)}{\sum_{i=\iota(s-1)}^{\iota(s)} \left(\Delta t_i^{\text{KMC}} n^l\right)} \tag{2.32}$$

With this the mean surface fraction of $\neg$B sites can be calculated as

$$\hat{\theta}_{\neg\mathrm{B}}(s) = 1 - \hat{\theta}_{\mathrm{B}}(s) \tag{2.33}$$

Using the average $\neg$B site fraction, the effective average fraction of the other species can be be determined, as given in the following for the example of the A^+ species:

$$\hat{\theta}_{\mathrm{A^+}} = \theta_{\mathrm{A^+(ads)},\neg\mathrm{B}} \cdot \hat{\theta}_{\neg\mathrm{B}} \tag{2.34}$$

This can be used to determine the effective reaction rate constants for forward and backward rates of process II and III, as shown in the following at the example of forward reaction of process II:

$$\hat{k}_{\mathrm{II}}^f(s) = \frac{\sum_{i=\iota(s-1)}^{\iota(s)} \Psi_i^1}{\Delta t_s^{\text{seq}} n^l \Delta L^2} \cdot \frac{1}{N_s \hat{\theta}_{\mathrm{A^+(ads)}} \exp\left(\frac{\beta\Delta\Phi\mathrm{F}}{RT}\right)}. \tag{2.35}$$

It is noted, that indeed, instead of reaction rate constants, one could also directly pass the reaction fluxes, e.g. q_{II}, from the kMC to the continuum model. However, for numerical stability of the solution it is advantageous to include some general dependencies within the continuum solution, if possible. For instance, here, the general dependency of the forward reaction of process II on surface fraction $\theta_{\mathrm{A^+(ads)}}$ is included within the continuum model, i.e. as given by Equation 2.7. Synchronization of the fluxes between kMC and continuum model is realized by adapting the rate constants k_{II}^{f} and k_{II}^{b} in the continuum model. Passing rate constants instead of fluxes, enables to impede, for instance, negative values of $\theta_{\mathrm{A^+(ads)}}$ in the continuum simulation and thus improves numerical stability, as stated above.

As mentioned above, kMC simulations are performed in parallel instances v. Thus, an output parameter $\hat{p}_{s,v}$, e.g. $\hat{\theta}_{\mathrm{B},s,v}$, depends on s and v. The mean value of the n^{par} parallel instances can thus be calculated as

$$M_s^p = \frac{\sum_{v=1}^{n^{\mathrm{par}}} \hat{p}_{s,v}}{n^{\mathrm{par}}}, \tag{2.36}$$

while the standard deviation of the calculated value M_s^p can be approximated as

$$S_s^p \approx \sqrt{\frac{1}{n^{\mathrm{par}}(n^{\mathrm{par}}-1)} \sum_{v=1}^{n^{\mathrm{par}}} |\hat{p}_{s,v} - M_s^p|^2} \tag{2.37}$$

It is noted that S_s^p, indeed, is only an approximation of the standard deviation, which is achieved by evaluating a sample of kMC simulations of the size n^{par}. In order to determine $p(t)$, e.g. $\theta_{\mathrm{B}}(t)$, for the continuum model a cubic smoothing spline is determined, that minimizes

$$\kappa \sum_{s=1}^{n^{\mathrm{seq}}} |M_s^p - p(\bar{t}_s^{\mathrm{seq}})| + (1-\kappa) \int \left| \frac{d^2 p(t)}{dt^2} \right|^2 dt \tag{2.38}$$

with smoothing factor κ. This yields the continuum input parameters as continuous functions of time t: $k_{\mathrm{II}}^{f}(t)$, $k_{\mathrm{II}}^{b}(t)$, $k_{\mathrm{III}}^{f}(t)$, $k_{\mathrm{III}}^{b}(t)$, and $\theta_{\mathrm{B}}(t)$.

2.2.5 Direct coupling algorithms

The previous section showed the continuum model, the kMC model, the sequencing and the exchanged values in detail. However, the order of passing values and the interpretation of simulation outputs have not yet been discussed. There are various options to realize this coupling which differ significantly in stability, accuracy and computation time. Here, a systematic analysis and comparison allows to show the pros and cons of the approaches, and thus educated selection and tailoring in future studies. Three different algorithms for coupling of both models are presented in this section.

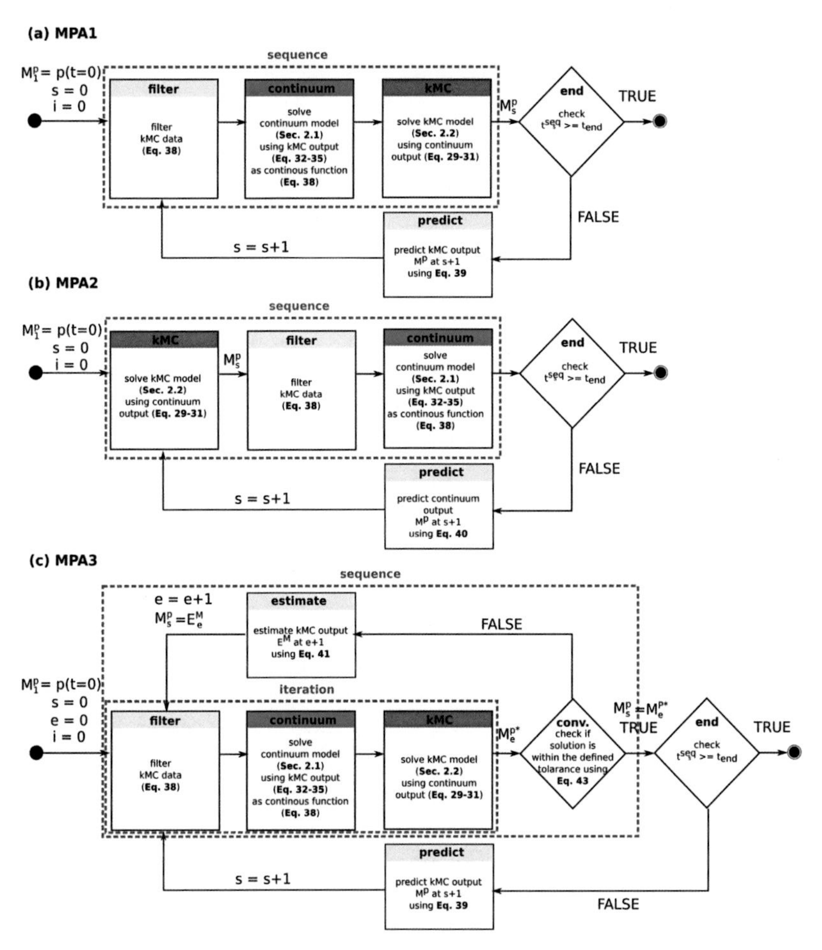

Figure 2.4: Illustration of the multiparadigm algorithms MPA1 (a), MPA2 (b) and MPA3 (c). Reprinted from publication [89] with permission from Elsevier.

In every algorithm within a sequence s, the kMC and the continuum model is solved. However, the order and the interpretation of the outputs can vary between algorithms. The first, MPA1, is shown in Figure 2.4a. Here, within a sequence, the continuum model is simulated first. In the second, MPA2, the order is switched, as can be seen in Figure 2.4b. The third, MPA3, as shown in Figure 2.4c, has the same sequential order as MPA1, but includes an estimation loop, which repeats one sequence s until the results are within a defined tolerance. As will be shown, the additional loop considerably increases computational cost of the overall simulation.

The algorithms include a filtering step of the kMC data to account for the stochastic nature of kMC, and a prediction step to predict the output of the second model in the upcoming sequence. This prediction step can cause an error in the overall simulation, which will be referred to as the *prediction error*. For MPA3, an estimation step is also introduced, in which the output of the kMC step is estimated based on the previous iterations. The intention is to correct the initially introduced prediction error. The simulation is terminated as soon as a specified end criteria is reached, which can be, for instance, $t > t^{\text{end}}$ with t^{end} being the specified end time.

In MPA1 and MPA3, the output of the kMC model for the upcoming sequence is predicted as

$$M^p_{s+1} = M^p_s + \frac{M^p_s - M^p_{s-1}}{\bar{t}^{\text{seq}}_s - \bar{t}^{\text{seq}}_{s-1}} (\bar{t}^{\text{seq}}_{s+1} - \bar{t}^{\text{seq}}_s) \tag{2.39}$$

in the upcoming sequence $p(t)$ is calculated based on Equation 2.38 including this prediction. In MPA2 the output of the continuum model is predicted as

$$p(\bar{t}^{\text{seq}}_{s+1}) = p(t^{\text{seq}}_s) + \frac{\mathrm{d}p(t^{\text{seq}}_s)}{\mathrm{d}t} \frac{\Delta t^{\text{seq}}_{s+1}}{2}. \tag{2.40}$$

Filtering of the kMC output is realized using the smoothing factor κ as given in Equation 2.38, while a smoothing factor of 1 yields no smoothing and smoothing factor of 0 yields an approximation by a linear function.

MPA3 includes an additional estimation loop, which is used to correct errors caused by the prediction step. The estimation procedure is provided in detail in the following. M^{p*}_e is the mean output of value p in estimation step e and E^M_e is the estimation of the the mean output. The estimation of the upcoming step $e+1$ is calculated as

$$E^M_{e+1} = E^M_e + K_P(M^{p*}_e - E^M_e) + K_I \sum_{z=1}^{e} (M^{p*}_z - E^M_z) \tag{2.41}$$

while the estimate of the standard deviation E^S_e is calculated accordingly, as

$$E^S_{e+1} = E^S_e + K_P(S^{p*}_e - E^S_e) + K_I \sum_{z=1}^{e} (S^{p*}_z - E^S_z) \tag{2.42}$$

Equations likewise consider the difference between the estimated value, e.g. E^M_e, and the actual output mean value, e.g. M^{p*}_e, in the previous step as well as the sum up to the last iteration step, which are thereby weighted by a proportional K_P and an integral factor K_I, respectively. This algorithm resembles the setup of a proportional-integral (PI) controller. Due to the fluctuations of the kMC output, E^M_e and M^{p*}_e

are unlikely to converge to exactly the same value. An acceptable tolerance between both values needs to be defined, which in this work is

$$E_e^M - E_e^S\lambda < M_e^{p*} < E_e^M + E_e^S\lambda \tag{2.43}$$

with λ being the tolerance factor to scale the tolerance relative to the standard deviation, i.e. fluctuations of the output. As soon as Equation 2.43 is true for all M_e^{p*}, the output of the sequence is defined as

$$M_s^p = M_e^{p*} \tag{2.44}$$

and the next sequence is calculated.

2.2.6 Error estimation

To estimate the error of the coupled simulation, results are compared to a pure continuum solution. This is possible because no heterogeneity is included in this kMC model. Nevertheless, the approach enables to consider heterogeneity in a coupled simulation has been shown in [100].

The kMC model covers the calculation of the reaction rate constants of reaction II and III, which for the homogeneous case can be determined as

$$k_{\mathrm{II}}^f = k_{\mathrm{II}} \exp\left(\frac{-E_{\mathrm{II}}^A}{RT}\right), \tag{2.45}$$

$$k_{\mathrm{II}}^b = k_{\mathrm{II}} \exp\left(\frac{-(E_{\mathrm{II}}^A - \Delta G_{\mathrm{II}}^0)}{RT}\right), \tag{2.46}$$

$$k_{\mathrm{III}}^f = k_{\mathrm{III}} \exp\left(\frac{-E_{\mathrm{III}}^A}{RT}\right), \tag{2.47}$$

and

$$k_{\mathrm{III}}^b = k_{\mathrm{III}} \exp\left(\frac{-(E_{\mathrm{III}}^A - \Delta G_{\mathrm{III}}^0)}{RT}\right). \tag{2.48}$$

Further, the model covers the balancing of species B on the adsorption site. In the purely continuum model this balance is included as

$$\frac{N_s}{o_s}\frac{\mathrm{d}\theta_{\mathrm{B}}(\mathrm{ads})}{\mathrm{d}t} = q_{\mathrm{II}} - q_{\mathrm{III}} \tag{2.49}$$

By supplementing the continuum model as provided in Section 2.1 with Equation 2.45-2.49, an accurate reference solution can be determined. In summary, this yields the following possible solutions for the simulation of the example problem:

- MPA1: Continuum model of Section 2.1. coupled with kMC model of Section 2.2. using MPA1 algorithm
- MPA2: Continuum model of Section 2.1. coupled with kMC model of Section 2.2. using MPA2 algorithm
- MPA3: Continuum model of Section 2.1. coupled with kMC model of Section 2.2. using MPA3 algorithm
- reference: Continuum model of Section 2.1 supplemented by Equations 2.45-2.49

To determine errors of the MPA algorithms, results are compared to the reference solution. Thus, absolute errors of the MPA solutions can be determined, which is shown for the example of the surface fraction of species B calculated by MPAx with x$\in \{1, 2, 3\}$ in the following:

$$\theta_{\mathrm{B}}^{\mathrm{error}}(\bar{t}_s^{\mathrm{seq}}) = \theta_{\mathrm{B}}^{\mathrm{MPAx}}(\bar{t}_s^{\mathrm{seq}}) - \theta_{\mathrm{B}}^{\mathrm{reference}}(\bar{t}_s^{\mathrm{seq}}). \tag{2.50}$$

Further, the absolute average error of potential, reaction rate constants, and surface fraction over all sequences are determined as

$$\bar{\Delta\Phi}^{\mathrm{error}} = \frac{\sum_{s=1}^{n^{\mathrm{seq}}} |\Delta\Phi^{\mathrm{error}}(\bar{t}_s^{\mathrm{seq}})|}{n^{\mathrm{seq}}}, \tag{2.51}$$

$$\begin{aligned}\bar{k}^{\mathrm{error}} &= \frac{\sum_{s=1}^{n^{s}} |k_{\mathrm{II}}^{f,\mathrm{error}}(\bar{t}_s^{\mathrm{seq}})|}{4n^{\mathrm{seq}}} + \frac{\sum_{s=1}^{n^{\mathrm{seq}}} |k_{\mathrm{II}}^{b,\mathrm{error}}(\bar{t}_s^{\mathrm{seq}})|}{4n^{\mathrm{seq}}} \\ &+ \frac{\sum_{s=1}^{n^{\mathrm{seq}}} |k_{\mathrm{III}}^{f,\mathrm{error}}(\bar{t}_s^{\mathrm{seq}})|}{4n^{\mathrm{seq}}} + \frac{\sum_{s=1}^{n^{\mathrm{seq}}} |k_{\mathrm{III}}^{b,\mathrm{error}}(\bar{t}_s^{\mathrm{seq}})|}{4n^{\mathrm{seq}}},\end{aligned} \tag{2.52}$$

and

$$\begin{aligned}\bar{\theta}^{\mathrm{error}} &= \frac{\sum_{s=1}^{n^{\mathrm{seq}}} |\theta_{\mathrm{A+}}^{\mathrm{error}}(\bar{t}_s^{\mathrm{seq}})|}{4n^{\mathrm{seq}}} + \frac{\sum_{s=1}^{n^{\mathrm{seq}}} |\theta_{\mathrm{B}}^{\mathrm{error}}(\bar{t}_s^{\mathrm{seq}})|}{4n^{\mathrm{seq}}} \\ &+ \frac{\sum_{s=1}^{n^{\mathrm{seq}}} |\theta_{\mathrm{C}}^{\mathrm{error}}(\bar{t}_s^{\mathrm{seq}})|}{4n^{\mathrm{seq}}} + \frac{\sum_{s=1}^{n^{\mathrm{seq}}} |\theta_{\mathrm{V}}^{\mathrm{error}}(\bar{t}_s^{\mathrm{seq}})|}{4n^{\mathrm{seq}}},\end{aligned} \tag{2.53}$$

respectively. Absolute errors for reaction constants for forward and backward reaction and those of the surface fractions are averaged; this allows to reduce the number of variables and thus enables a focused discussion and comparison of the MPAs.

The errors, as introduced above, are used to evaluate the overall error of the coupled simulation. However, those evaluations barely allow to assign errors to certain causes. Most relevant causes are the fluctuation of the kMC (fluctuation error), the quasi steady state assumption within a kMC sequence neglecting the transient change of the input parameter (transition error), and the prediction of the input values (prediction error). To decompose the effect of those contributions, they are investigated separately, as will be outlined in the following. By assuming constant values for the input of the kMC model, as given in Equation 2.29-2.31, an error is introduced, which will be referred to as transition error. The error can be decomposed by using the according assumption in the reference solution. In the following, this is shown for the example of the electrical potential $\Delta\Phi$, as used used in Equation 2.7:

$$\Delta\Phi(t) = \Delta\Phi(\bar{t}_s^{\mathrm{seq}}) \tag{2.54}$$

for $t_{s-1}^{\mathrm{seq}} < t \leq t_s^{\mathrm{seq}}$. According to Equation 2.40 the error of MPA2, which includes the prediction, can be introduced at the example of $\Delta\Phi$ by assuming

$$\Delta\Phi(t) = \Delta\Phi(t_{s-1}^{\mathrm{seq}}) + \frac{\mathrm{d}\Delta\Phi(t_{s-1}^{\mathrm{seq}})}{\mathrm{d}t}\frac{\Delta t_s^{\mathrm{seq}}}{2} \tag{2.55}$$

According to Equation 2.39 the error of MPA1, which includes the prediction, can be introduced at the example of θ_{B} by assuming

$$M_s^{\theta_{\mathrm{B}}} = M_{s-1}^{\theta_{\mathrm{B}}} + \frac{M_{s-1}^{\theta_{\mathrm{B}}} - M_s^{\theta_{\mathrm{B}}}}{\bar{t}_{s-1}^{\mathrm{seq}} - \bar{t}_s^{\mathrm{seq}}}(\bar{t}_s^{\mathrm{seq}} - \bar{t}_{s-1}^{\mathrm{seq}}) \tag{2.56}$$

for $t_{s-1}^{\mathrm{seq}} < t \leq t_s^{\mathrm{seq}}$ and application of the smoothing function in Equation 2.38 to determine $\theta_{\mathrm{B}}(t)$. The assumptions that are given in Equations 2.55 and 2.56, however, also include a transition error, which need to be subtracted to decompose the error of the prediction. The errors caused by the fluctuation of the kMC output can be quantified by the standard deviation of the kMC output, i.e. Equation 2.37. To sum up, the particular error contributions in this work have been estimated as absolute errors, which are calculated as difference between the reference model and the reference model with modifications outlined above. With this, the contributions to the overall error of the three causes listed above, are decomposed for a certain value, e.g. the surface fraction of B, as given in the following:

- transition error of value = value in reference model modified by assumptions according to Equation 2.54 − value in reference model
- prediction error of value (MPA1) = value in reference model modified by assumptions according to Equation 2.55 − transition error − value in reference model
- prediction error of value (MPA2) = value in reference model modified by as-

sumptions according to Equation 2.56 – transition error – value in reference model

- fluctuation error of value = S_s^p in Equation 2.37

2.2.7 Simulation scenario and parameters

The input into the simulation triggering the reaction is an externally applied current I. The current leads also to charge and discharge of the electrochemical double layer and change of the electrical potential $\Delta\Phi$ at the electrochemical surface. Since the presented approach is designed to be applied to investigate dynamic operation of electrochemical systems, also a dynamic signal is applied. A sinusoidal input current is defined as

$$I = \bar{A} \sin(2\pi f t) \tag{2.57}$$

with amplitude $\bar{A}$ and the frequency f. All parameters applied in this chapter are listed in the appendix A in Table A.1. Parameters are chosen in a physically reasonable order of magnitude but mainly to provide illustrative simulation results. All models are implemented and solved in MATLAB. Ordinary differential equations are solved by the ode15s solver. Simulations are performed in Matlab on a 64-bit linux system with Intel Core™ i7-3770 CPU with 3.40 GHz $\times$ 8 and 15.5 GB RAM.

2.3 Results and discussion

To show the impact of the type of coupling algorithm and configuration, simulations were performed with a varying number of lattice sites n^l, number of sequences n^{seq}, proportional factor K_P, tolerance factor λ, and smoothing factor κ. Multiparadigm simulations are benchmarked by comparison to a pure continuum solution, indicated as reference. The standard configuration of the algorithms is defined by $n^l = 10^2$, $n^{\text{par}} = 16$, $K_P = 0.2$, $\lambda = 1$, and $\kappa = 1$ and is always applied if not stated otherwise.

2.3.1 Numerical accuracy of the coupling algorithms

Figure 3.3 shows the electrical potential $\Delta\Phi$ for the continuum model and for the multiparadigm algorithms MPA1, MPA2, and MPA3 using the reference configuration. Due to the sinusoidal signal input of the external current I, the electrical potential increases and decreases as a sinusoid also. All three algorithms are in good agreement with the continuum solution with deviations below 5% relative to the voltage amplitude. The errors of MPA1 and MPA2 are significantly higher than MPA3. Further,

MPA1 and MPA2 both possess an oscillatory increase and decrease of the error. The peaks of the errors are increasing for both algorithms towards the end of the simulation and are comparable regarding positions in time and magnitude, which suggests both errors are of the same origin. In contrast, the error of MPA3 neither increases nor shows a comparable systematic behavior. The electrical potential is determined by the continuum part of the MPAs with Equation 2.1, and depends on the current input I and the rate of reaction II, q_{II}. Since the input I is equivalent for all simulations, the error of the electrical potential originates from an error of the reaction rate q_{II}. In the MPA simulations, this rate directly depends on three kMC outputs, being the forward and backward reaction rate constants, k_{II}^{f} and k_{II}^{b}, and the surface fraction of species B, θ_{B}. Although the simulation accuracy is considerably higher for MPA3 compared to MPA1 and MPA2, its computational cost is nearly 30 times higher, as given in Figure 3.3.

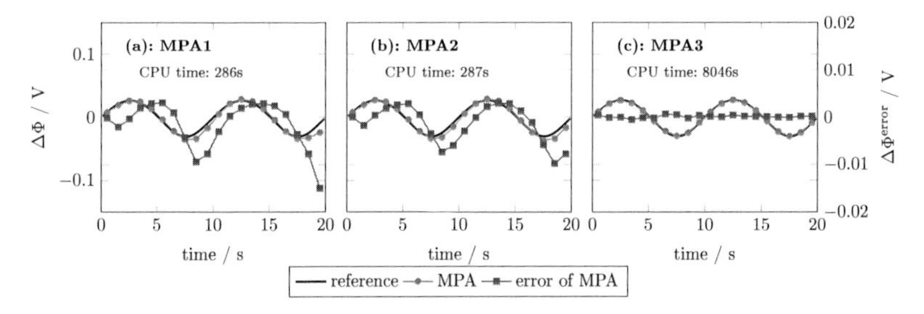

Figure 2.5: Difference in the electrical potential $\Delta\Phi$ and absolute error compared to the continuum solution, and the CPU time for algorithms MPA1, MPA2, and MPA3. Reprinted from publication [89] with permission from Elsevier.

While the errors for MPA1 and MPA2 have the same systematic behavior, MPA1 had a larger maximum error and so was analyzed in more detail. In Figure 2.6, the concentrations of species A^{+} and C in solution are compared for the continuum-only model and MPA1. Just as for the electrical potential, the concentrations are solved within the continuum part of the MPAs, and errors indirectly originate from errors in the kMC output. The concentration of species A^{+} is in very good agreement with the continuum solution, whereas the concentration of species C has a considerable deviation, as the absolute error of the concentration is about two orders of magnitude higher compared to the absolute error of concentration of species A^{+}. Both errors show the same systematic behavior as previously seen for the electrical potential. The concentrations are influenced by surface fractions of species A^{+}, C, and vacancies V, so the surface fractions and their errors are discussed next.

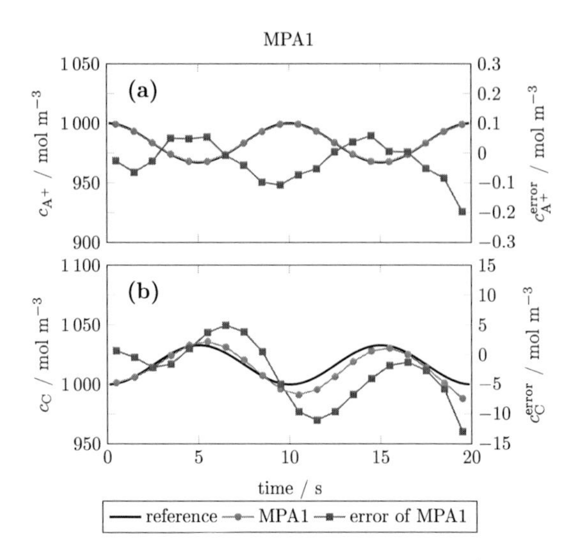

Figure 2.6: Concentration and absolute error of species (a) A^+ and (b) C for MPA1. Reprinted from publication [89] with permission from Elsevier.

In Figure 2.7, MPA1 is further evaluated for the surface fractions θ_i of species A^+, B, C, and vacancies V. Again, the impact of the sinusoidal input can be seen, as the surface fractions decrease and increase sinusoidally as well. The amplitude of the oscillations is lowest for species A^+ and V and highest for species B. The mean and amplitude of the surface fraction of vacant sites and of species A^+ are very similar. The deviations of the surface fractions in MPA1 are highest for species B and lowest for species A^+ and vacancies V, which is shown in detail in Figure 2.8. The errors in the surface fractions originate from errors in the kMC outputs and in particular directly from error in the surface fraction θ_B. Since surface fractions of all species sum up to 1 as given by Equation 2.10, the absolute error in the surface fractions sum up to 0. As such, an error in the surface fraction of species B forces a distribution of this error on the other surface fractions. The share of this error is almost equal for species A^+ and vacancies V. The very low observed error in the concentration of A^+ in the solution can be explained. Comparing to the continuum solution, the surface fractions of the vacancies V and of species A^+ impact the forward and backward rate of process I, respectively. Since both surface fractions have almost the same values and errors, the impacts compensate each other leading to a very low error propagation to the concentration of A^+ in the solution. In contrast, for reaction 4, the values and errors of the surface fraction of C considerably differ leading to a distinct and significant error propagation to the solution concentration C. The results illustrate that the propagation of an error in a multiscale algorithm can depend considerably on the system and its parameters.

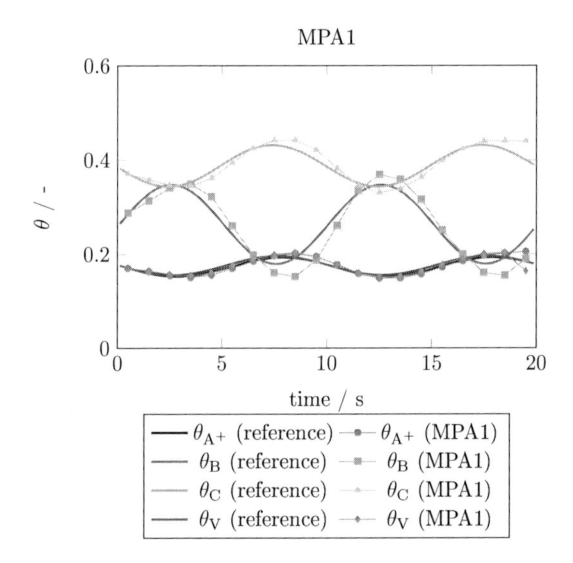

Figure 2.7: Surface fraction of species A^+, B, C, and vacancies V for the continuum solution and MPA1. Reprinted from publication [89] with permission from Elsevier.

The magnitude and systematic nature of the errors of MPA1 and MPA2 (not shown here in detail) were found to be comparable, and the order of the sequence was of minor relevance for the simulation accuracy and computational cost. MPA3 significantly improved the accuracy but with much higher computational cost. The systematic oscillatory nature of the errors in MPA1 and MPA2, which do not appear in MPA3, suggests a different origin. The next section analyzes the errors in detail and discusses efficient tuning of the algorithms to reduce the overall simulation error.

2.3.2 Algorithm tuning

By examining the system states (i.e. potentials, concentrations, and surface fractions), the impact of MPAs on the simulation error has been shown. Errors can originate thereby from the coupling of the continuum and kMC models, as well as the stochastic nature of the kMC model. Different algorithms can lead to very different systematic errors and magnitudes of the errors. In particular, a considerable difference between algorithms without estimation correction, i.e. MPA1 and MPA2, and an algorithm in-

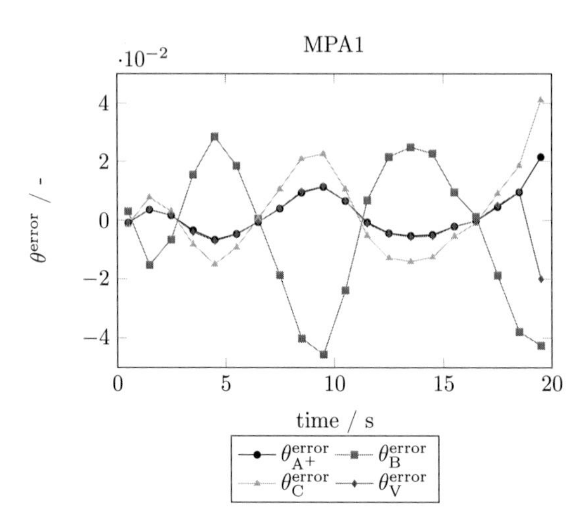

Figure 2.8: Absolute error of surface fraction of species A^+, B, C, and vacancies V for MPA1. Reprinted from publication [89] with permission from Elsevier.

cluding an estimation correction, i.e. MPA3, could be seen. As introduced above,three types of errors are distinguished: (i) prediction error, (ii) transition error, and (iii) fluctuation error. The errors of the reaction rate constants were not dependent on time and so do not include any prediction or transition errors. As such, only the results for the surface fraction θ_{B} are discussed in detail.

The various absolute errors of θ_{B} for the MPAs are shown in Figure 2.9. The prediction and transition errors are oscillatory. The fluctuation error induced by the kMC simulation is nearly constant during the whole simulation. The prediction error is considerably larger than the transition and fluctuation errors, which are of the same order of magnitude. The overall error for MPA1 is oscillatory and strongly correlates with the prediction error, especially during the first quarter of the simulation; as the simulation proceeds, the overall error of the MPA1 increases and exceeds the prediction error. The errors accumulate and the MPA solution becomes somewhat out of phase with ongoing simulation time. A similar correlation between the prediction error and overall error occurs for MPA2. The prediction error of MPA2 is slightly lower, with some increase in peak overall error with time. For MPA3, the initial prediction is equal to that for MPA1, but is corrected by the estimation correction loop and so is not seen in the 2.9c. The estimation correction leads to a considerable decrease of the overall error.

The fluctuation error can be reduced by increasing the number of lattice sites, such as increasing the grid size. Figure 2.9 indicates that increasing n^l from 10^2 to

20^2 did not reduce the overall error of MPA1 and MPA2, which is dominated by the prediction error, but did reduce the overall error for MPA3. The reduction in fluctuation error for MPA3 results in its overall error becoming more correlated to the transition error. These results illustrate that increasing the number of lattice sites can only be an efficient strategy if the fluctuation error is the dominant, because of the overall simulation error.

The computational cost for the simulations with an increased field size are also provided in Figure 2.9. The computational cost is significantly increased for all simulations compared to that of $n^l = 10^2$, given in Figure 3.3. Again, the computational cost of MPA3 is significantly higher, i.e. more than 30 times higher, compared to MPA1 and MPA2. As such, there is a strong incentive to avoid a computationally expensive estimation-correction loop or to significantly reduce its iterations.

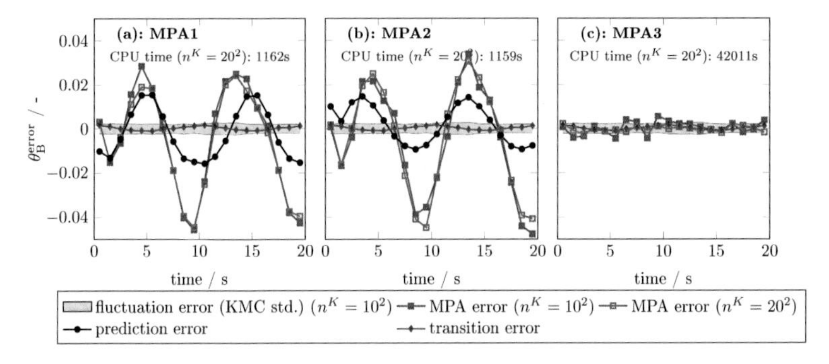

Figure 2.9: Errors for (a) MPA1, (b) MPA2, and (c) MPA3 for surface fraction of species B for $n^l = 10^2$ and $n^l = 20^2$, as well as prediction, transition, and fluctuation errors for $n^l = 10^2$. The CPU time for $n^l = 20^2$ is given in each subplot. Reprinted from publication [89] with permission from Elsevier.

A systematic analysis of several measures to tune the MPA is shown by varying the field size n^l in Figure 2.10adg, the smoothing factor κ in Figure 2.10beh, and the number of sequences n^{seq} in Figure 2.10cfi. The results show the effect of aforementioned measures on the averaged absolute errors of the electrical potential $\Delta\Phi$ in Figure 2.10abc, the reaction rate constants k^f_{II}, k^b_{II}, k^f_{III} and k^b_{III} in Figure 2.10def, and the surface fractions $\theta_{\text{A+}}$, θ_{B}, θ_{C}, and θ_{V} in Figure 2.10ghi for the three MPAs. The electrical potential can be interpretted as being an overall simulation error, as it represents the typical electrochemical model output. The reaction rates are kMC output variables that are independent of time and thus are only affected by fluctuation errors. The surface fractions are variables that are additionally directly affected by transition and prediction errors.

In Figure 2.10g, the errors in surface fractions are shown for a varying field size. The error is not affected by n^l for MPA1 and MPA2 since their errors are dominated by prediction error, but the error can be reduced for MPA3, which has no prediction error. Prediction errors are not affected by the number of lattice sites, which here makes increasing the field size a futile measure. In contrast, errors in the reaction rate constants, as shown in Figure 2.10d, are reduced for increasing the field size for all three algorithms. This error is only affected by fluctuation errors, which can be efficiently reduced by increasing the number of lattice sites. In Figure 2.10a, errors in the electrical potential is shown. Again, errors do not possess a clear dependency on field size for MPA1 and MPA2. In contrast, the errors of MPA3 can be slightly decreased with increased n^l. The error in the electrical potential is more strongly influenced by the error of the surface fraction than by the error of the reaction rate constant. For the variation in field size in Figure 2.10, the error in MPA3 is an order of magnitude lower for electrical potential and surface fraction than for MPA1 and MPA2, but only about 40% lower for the reaction rate constant. The accuracy can be improved mostly by an estimation correction if the error is dominated by prediction error.

Next, consider the effects of smoothing the kMC output. Smoothing the kMC output, i.e. decreasing κ, reduced the overall error dominated by fluctuations of the kMC model (Figure 2.10e). In contrast, smoothing had minimal effects on the overall error dominated by prediction error, i.e. the error in surface fraction θ for MPA1 and MPA2 (Figure 2.10h), and increased smoothing increased the overall error for MPA3, which has no prediction error. The trends in the errors in the electrical potential are similar to those for θ (Figure 2.10b), as those errors are more closely related than the errors in reaction rate constants in this simulation scenario.

Finally, consider the impact of varying the number of sequences, n^{seq}. An increasing number of sequences corresponds to a reduction of the time length of each sequence. Shorter time steps will reduce prediction length and improve accuracy of mean approximations within a kMC simulation, which will reduce the prediction and transition errors. However, shorter time sequences will provide less data from the kMC model and thus result in an increase in the fluctuation error. The consequences can be observed in Figure 2.10cfi. The errors in surface fraction θ for MPA1 and MPA2 in Figure 2.10i, which are dominated by the prediction error, can be significantly reduced to nearly the same magnitude as the error of MPA3 with an increasing number of sequences ,n^{seq}, from 10 to 50. Further increase in the number of sequences increases the error.

In contrast, the error in the reaction rate constants for all MPAs increase with increasing number of sequences (Figure 2.10f), as this error is dominated by kMC fluctuations, whose effects are increased by lowering the number of kMC steps. The trends in the errors for the electrical potential in Figure 2.10c mostly mirror the errors for the surface fraction θ, except for stronger error reduction for MPA3 when decreasing the number of sequences, due to reduced fluctuation errors.

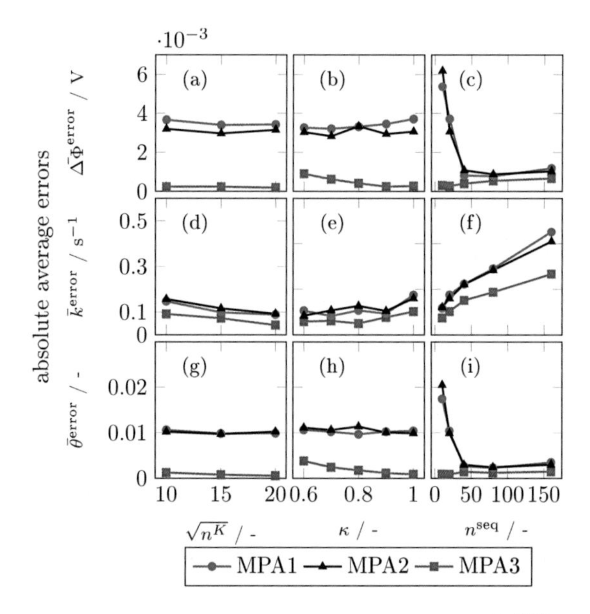

Figure 2.10: Absolute average errors for the electrical potential $\bar{\Delta\Phi}^{\text{error}}$, reaction rates $\bar{k}^{\text{error}}$, and surface fractions $\bar{\theta}^{\text{error}}$ for varying field size, smoothing factor, and number of sequences. Reprinted from publication [89] with permission from Elsevier.

In general, the trend and magnitude of the errors for MPA1 and MPA2 are comparable for variations, and confirm that the sequential order of the stochastic and continuum models was of minor relevance. Distinct effects could be observed between algorithms with and without estimation correction and in particular for the number of sequences and smoothing of the kMC output. In general, the accuracy of MPA3 was the highest for all simulations and configurations shown here. The high accuracy was achieved by correcting the prediction error but with significantly increased computational cost. In summary, these results clearly examined the origin of the errors of the MPAs for the simulation scenario. The overall error was dominated by the largest of the three types of error (prediction, transition, fluctuation), and a reduction of less dominant errors did not reduce the overall error of the simulation.

2.3.3 Efficient estimation correction

The estimation-correction loop in MPA3 evaluates the kMC output and iterates until the kMC output is within a defined tolerance region. The definition of the tolerance and the approach to control the solution into this tolerance region determines the accuracy and computational cost of MPA3. The computational cost is strongly related to the number of iterations needed in each time interval. To enable cost-effective multiscale simulations, therefore, it is important to identify a configuration of the algorithm that minimizes the number of iterations. This section analyzes the impact of n^l, K_P, and λ.

Figure 2.11 shows convergence results for the reference configuration, which takes about 450 iterations to reach $t^{\mathrm{end}} = 20$ s. The reaction rate constants are within tolerance for most of the iterations, because the parameters are independent of time and the estimation-correction iterations are not needed (Figure 2.11a). Nevertheless, an output can be outside the specified tolerance because of stochastic fluctuations. In this simulation, only the kMC output θ_{B} changes due to its dependency on the applied sinusoidal current input. Therefore, θ_{B} is most critical and its progress over the iterations is shown in detail in Figure 2.11b. Within one sequence, the estimated value approaches the actual value until it is within the specified tolerance. The number of iterations needed can considerably deviate between sequences. The probability that one of the evaluated parameters is outside the specified tolerance because of stochastic fluctuations increases with the number of evaluated parameters. As the algorithm needs to hold all the selected output parameters within the specified tolerance, convergence can be challenging when having more than one interdependent output. The problem as a whole is similar to a robust control problem [103].

Figure 2.12 shows convergence of the critical variable θ_{B} for varying simulation parameters. Increasing the field size decreases the fluctuations of the kMC output (cf. Figures 2.11b and 2.12a), which decreases the output uncertainty and in general should help to control the tolerance range. However, as tolerance is defined as a function of standard deviation, as given with Equation 2.43, the total number of iterations increased to almost 600 iterations. Moreover, the computational cost of a single iteration significantly increases due to increased number of grid elements, resulting in a total computational time of 42,011 s instead of 8046 s. This factor of five increase in computational cost only produces a slightly improved accuracy (Figure 2.10a). As discussed in the previous section, the overall error of the solution cannot be reduced below the transition error, which is not affected by increasing the field size. Further increasing the field size will thus not further reduce the error of θ_{B}. Considering the significantly increased computational cost, the slight improvement of accuracy is not worth the effort for this example.

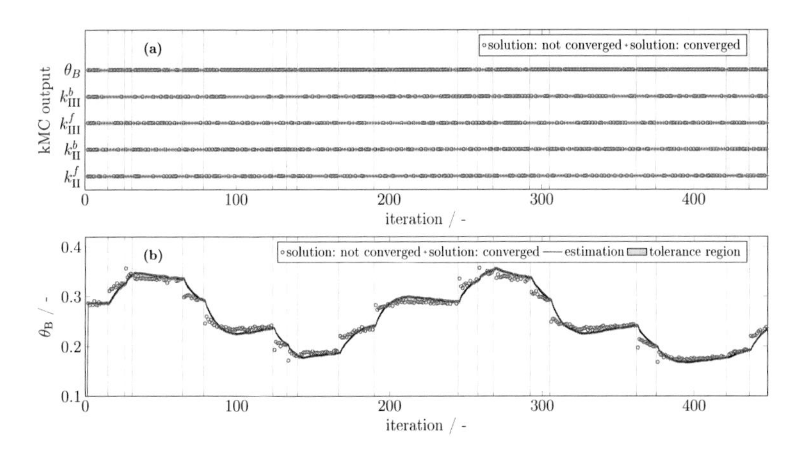

Figure 2.11: (a) Convergence status during the iteration process and (b) estimated value, tolerance region, and actual kMC output for the surface coverage of B, θ_B. The convergence status of the kMC outputs are indicated as converged (blue), i.e. output is within the tolerance region, and not converged (red), i.e. output is outside the tolerance region. End time of a sequence is indicated by vertical gray lines with the last time step corresponding to t^{end}. This reference configuration has $n^l = 10^2$, $n_{instance} = 16$, $K_P = 0.2$, $\lambda = 1$, and $\kappa = 1$. Reprinted from publication [89] with permission from Elsevier.

Increasing the proportional factor K_P from 0.2 to 0.8 gives faster convergence (Figure 2.12b), with a considerable reduction of the number of iterations compared to the reference configuration. Drawbacks of the higher proportional factor are overshooting and oscillation of the estimation. A systematic optimization of K_P and K_I would lead to a faster convergence without oscillation, i.e. a further reduction of the number of iterations and computational cost.

Decreasing the tolerance factor λ from 1 to 0.75 increases the number of iterations significantly, from 450 to more than 800 iterations (Figure 2.12c). In several sequences, the estimation reaches a steady value, but the tolerance region is not reached, which suggests that the chosen tolerance is too low compared to the fluctuation of the kMC output. Although the estimation is good, the kMC output variables are often outside the tolerance region. One option for reducing the iterations is to define the tolerance according to the standard deviation or higher.

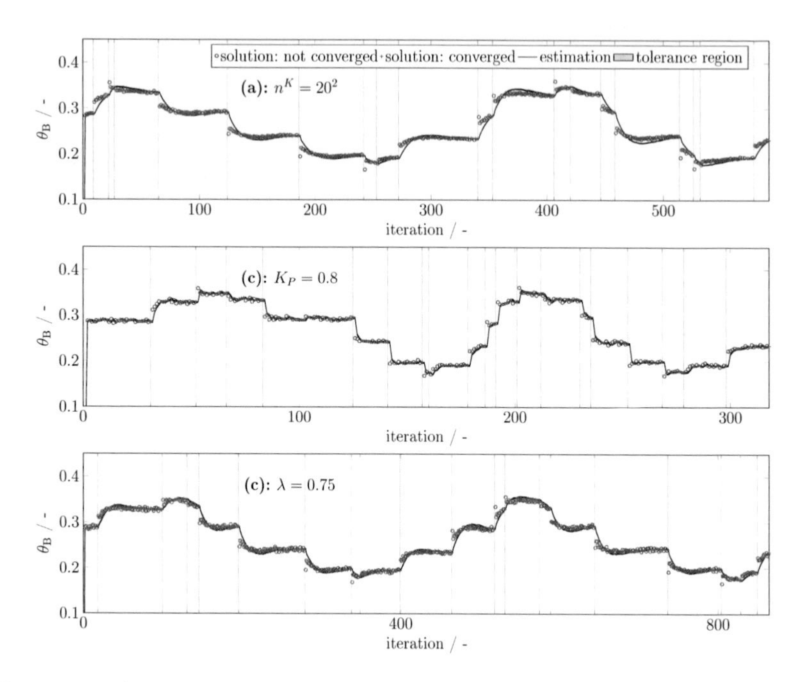

Figure 2.12: Convergence behavior for the kMC output θ_B for variations of the algorithm parameter set: (a) increased field size, (b) increased proportional factor, and (c) decreased tolerance factor. Reprinted from publication [89] with permission from Elsevier.

2.4 Concluding remarks

In this chapter a systematic analysis of coupling algorithms for the multiscale simulation, i.e. coupling of kMC and continuum models, has been presented. Any of three presented algorithms can be a good choice for multiparadigm simulation, depending on the needs and tradeoffs in numerical accuracy and computational cost. The coupling algorithms without estimation-correction were more than an order-of-magnitude less computationally expensive, but also an order-of-magnitude less accurate. Errors in the coupled simulations originate from several different causes and are categorized into fluctuation, prediction, and transition errors. Measures that are efficient in reducing one error type may be futile or even counterproductive regarding the other error types. Quantitative analysis as done in this study should be carried out to learn the underlying causes of numerical errors to facilitate an efficient and robust design of the coupling algorithm. While computationally efficient algorithms such as MPA1 and MPA2 can be suitable for some applications, the sequence size length needs to be considered carefully to keep fluctuation errors and sequence size-dependent errors in the same order of magnitude. Algorithms that include estimation-correction, such as MPA3, are much more robust regarding configuration and possess the highest accuracy for all configurations evaluated here.

This chapter provides a guide for the systematic selection and design of computationally efficient and accurate coupling simulations. Based on the findings, inappropriate or inefficient multiscale simulations can be prevented in the following chapters, where developed algorithms are applied to study surface degradation processes in lithium-ion batteries.

Chapter 3

Multiparadigm Simulation of Heterogeneous Film Growth Mechanism[4]

3.1 Introduction

In this chapter, the multiparadigm method shown in chapter 2, is used to investigate heterogeneous surface film growth mechanisms in lithium-ion batteries. The focus is thereby laid on testing the method at a plausible case study. Therefore, a mechanism for decomposition of Ethylene Carbonate (EC), i.e. the most commonly used solvent, on a graphite particle is implemented based on literature findings. The chapter thereby discusses both, numerical aspects of coupling kMC and continuum code at this actual use case and the physical interpretation of simulation results.

Degradation of lithium-ion batteries in context of the SEI has been inroduced in chapter 1. It is the most important but still not well understood ageing phenomenon at graphite negative electrodes [107, 108, 109]. Graphite is the common negative electrode and operates at conditions outside the electrochemical stability window of the electrolyte [108, 110]. Electrolyte decomposition takes place at the surface of the electrode particles and leads to the formation of a surface film, i.e. solid electrolyte interface (SEI). The SEI is mainly built during the first cycles prior to use and is considered to be part of the manufacturing process [111]. The aim of the formation process is to create an interface that is a good lithium-ion conductor, i.e. allows battery operation, but that is insulating for electrons and prohibits direct contact between electrode and electrolyte in order to provide good performance and long lifetime [110]. Different compositions of the film have been proposed by different research groups [76, 112]. Since the composition and structure and so the film characteristic

[4]Part of this chapter has been published in (Röder et al., J. Electrochem. Soc., 164(11), E3335-E3344, 2017 [106])

is determined during the first cycles [112], a detailed understanding of this growth mechanism is needed to improve cycling performance.

The SEI is formed by a complex mechanism [111]. An atomistic reaction mechanism involves lithium salt and solvent as reactants as well as a variety of different organic and inorganic intermediates and solid products [112, 113]. The observation of the formation process is challenging, because of the film's thickness of only several nanometers. Additionally, macroscopic properties such as particle size or operating conditions, e.g. C-rate and environmental temperature, have an important impact on the formation process [110, 109]. Although SEI film formation has been studied for decades using experimental and simulation-based methods, the exact mechanism of chemical and electrochemical reactions and the growth of the solid film is not understood. Indeed, Kalz et al. [91] pointed out the general need for a deeper analysis and advanced modeling of changes of reactive surfaces. New modeling methods need to be explored which can be applied to lithium-ion batteries for a detailed simulation of heterogeneous growth mechanisms, while considering the experimentally determined impact on macroscopic properties.

Several model-based approaches have been proposed for studying the SEI. Models are available at different scales based on molecular dynamics (MD), density functional theory (DFT), kinetic Monte Carlo (kMC), and partial differential equations (PDEs). MD simulations have been used to determine and analyze basic atomistic processes [76, 77, 114] and usually assume ideal surfaces or solutions. DFT has been applied to determine energy barriers and standard state potentials [80] and transport processes in the solid film [84, 83]. Methekar et al. [72] proposed the application of kMC to simulate heterogeneous passivation of the interface. Several macroscopic electrode models are available based on PDEs to simulate film growth and resistance [33, 115] and for detailed analysis of transport processes and reactions in the SEI [30]. Most atomistic methods can only be applied on very short time and length scales and cannot simulate the long time scale formation process in an electrode. On the other hand, homogeneous macroscopic methods do not consider process heterogeneity due to lateral interaction of species on a higher dimensional surface, which results in the complex structures observed experimentally. Using kMC method in combination with macroscopic continuum equations has been shown to be a promising approach for analysis of multiscale problems in other electrochemical systems, such as fuel cells [99] and copper electrodeposition [86]. To enable understanding the complex film formation mechanisms in lithium-ion batteries, in this chapter the methodology for direct multiparadigm simulation of chapter 2 is applied to couple heterogeneous surface reactions and film growth mechanisms with a macroscopic single-particle (SP) electrode model.

After explaining the approach, simulations are qualitatively validated for key aspects of the formation process, such as the SEI formation plateau. The method is demonstrated by simulating the formation process for an electrolyte with a pure EC solvent, which is known to form a stable film on the surface. The scope of this chapter is not to fully validate or parameterize the model, but to introduce the computational framework for the testing of hypothesized mechanisms that can be compared with experimental data and used for multiscale analysis, which is done in detail in chapter 4.

3.2 Multiparadigm model of film growth

A macroscopic electrode model based on mass and charge balances is dynamically coupled with a kinetic Monte Carlo model covering heterogeneous surface film growth mechanisms. The individual models can also be used without the coupling algorithm and have been evaluated separately in preliminary steps. Algorithms and mathematical models are described in detail in this section. The multiscale model and the various phases considered are illustrated in Figure 3.1.

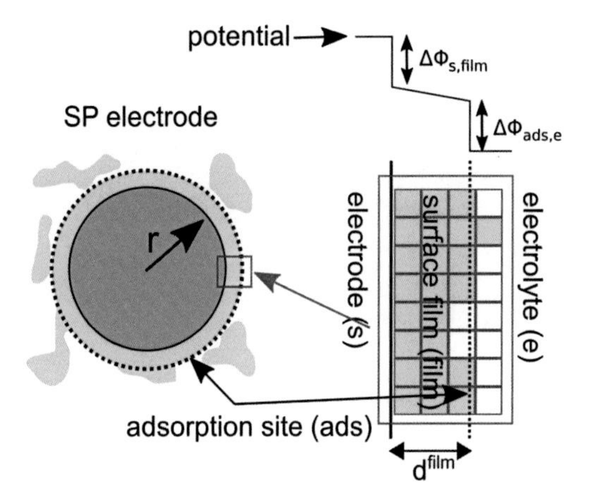

Figure 3.1: Multiscale model showing an electrode represented as a SP covered by a dense inner surface film (film) and further solids in the electrolyte reducing the porosity. A zoom into the film structure is indicated by a red square. Further, electric potential drop from electrode (s) through surface film (film) and adsorption site (ads) to electrolyte (e) is illustrated. Reprinted from publication [106].

3.2.1 Macroscopic scale

The macroscopic model considers mass and charge conservation, transport processes through SEI and active material particles, as well as kinetic limitations, i.e., charge transfer reactions. All phases shown in Figure 3.1 are involved, which are: electrode (s), surface film (film), adsorption site (ads), and electrolyte (e).

The model is based on a SP model approach similar to that of Santhanagopalan et al. [41] and is extended by SEI based on the work of Colclasure et al. [30]. Diffusion of lithium in the solid phase of the negative electrode, i.e. anode, is modeled as

$$\frac{\partial c_s^{\mathrm{Li(s)}}(r)}{\partial t} = \frac{1}{r^2}\nabla(D_s^{\mathrm{Li(s)}} r^2 \nabla c_s^{\mathrm{Li(s)}}(r)) \tag{3.1}$$

where $c_s^{\mathrm{Li(s)}}$ is the lithium concentration, r the radial dimension, and $D_s^{\mathrm{Li(s)}}$ the solid diffusion coefficient in the anode. The boundary conditions are $-D_s^{\mathrm{Li(s)}}\nabla c_s^{\mathrm{Li(s)}}(0) = 0$ and $-D_s^{\mathrm{Li(s)}}\nabla c_s^{\mathrm{Li(s)}}(R_s) = q_{\mathrm{s,film}}^{\mathrm{Li}} r_s$, with particle radius R_s and lithium flux $q_{\mathrm{s,film}}^{\mathrm{Li}}$ at the interface between anode and surface film and surface roughness factor r_s. The SEI is a surface film, which encloses the active material particle and is conductive only for lithium ions. The surface film is assumed to be planar, since its thickness is orders of magnitude smaller than that of the particle radius. Further, the SEI is treated as a single ion conductor with assumed electroneutrality, which denotes constant lithium concentration in the SEI and enforces equal flow rates for lithium through the s/film and film/ads interfaces. Therefore, reactions for lithium passing those interfaces are treated as a homogeneous reaction in series as explained by Helfreich [116], which allows the elimination of intermediate concentrations in the SEI phase, i.e. film, and determination of surface fluxes $q_{\mathrm{s,film}}^{\mathrm{Li}}$ and $q_{\mathrm{film,ads}}^{\mathrm{Li^+}}$ from phase s to phase film as

$$q_{\mathrm{s,film}}^{\mathrm{Li}} = q_{\mathrm{film,ads}}^{\mathrm{Li^+}} = \frac{N_s}{o_s}\left(\frac{a^{\mathrm{Li(s)}}(R_s)\lambda_{\mathrm{s,film}}\lambda_{\mathrm{film,ads}}}{\lambda_{\mathrm{film,s}}+\lambda_{\mathrm{film,ads}}} - \frac{\theta^{\mathrm{Li^+(ads)}}\lambda_{\mathrm{film,s}}\lambda_{\mathrm{ads,film}}}{\lambda_{\mathrm{film,s}}+\lambda_{\mathrm{film,ads}}}\right) \tag{3.2}$$

$$\lambda_{\mathrm{s,film}} = k_{10}^f \exp\left(\frac{-E_{10}^A + \beta_{10}\Delta\Phi_{\mathrm{s,film}}\mathrm{F}}{\mathrm{R}T}\right) \tag{3.3}$$

$$\lambda_{\mathrm{film,s}} = k_{10}^b a^{\mathrm{V(s)}}(R_s) \exp\left(\frac{-E_{10}^A - (1-\beta_{10})\Delta\Phi_{\mathrm{s,film}}\mathrm{F}}{\mathrm{R}T}\right) \tag{3.4}$$

$$\lambda_{\mathrm{film,ads}} = k_{11}^f \theta^{\mathrm{V(ads)}} \exp\left(\frac{-E_{11}^A}{\mathrm{R}T}\right) \tag{3.5}$$

$$\lambda_{\mathrm{ads,film}} = k_{11}^b \exp\left(\frac{-E_{11}^A}{\mathrm{R}T}\right) \tag{3.6}$$

with surface fraction $\theta^{\mathrm{Li^+ads}}$ of lithium, site occupancy number o_s, site density N_s, symmetry factor β, and activity of lithium $a^{\mathrm{Li(s)}}$ and vacancies $a_a^{\mathrm{V(s)}}$ in the anode, which are determined as a function of concentration $c^{\mathrm{Li^+(s)}}$ based on Redlich Kister coefficients for graphite provided by Colclasure et al. [117].

At the adsorption site, ads, of the film, balance equations for considered species m are applied as

$$\frac{N_s}{o_s}\frac{\partial\theta^m}{\partial t} = q^m_{\mathrm{film,ads}} + q^m_{ads,e} + \sum \nu^m Q^m_{\mathrm{ads}} \tag{3.7}$$

with fluxes through the interface q^m, source terms through reactions on the adsorption site Q^m, and stoichiometric factor ν^m. Those terms can be specified by continuum equations as shown by Colcasure et al. [30] or be a result of atomistic simulations as given in the multiscale coupling section.

Charge balance equations are applied for the s/film interface and the ads/e interface as

$$a_s^{\mathrm{eff}} C^{\mathrm{DL}}_{\mathrm{s,film}} \frac{\partial \Delta\Phi_{\mathrm{s,film}}}{\partial t} = -j^{\mathrm{charge}} + j^{\mathrm{ct}}_{\mathrm{s,film}} \tag{3.8}$$

and

$$a_s^{\mathrm{eff}} C^{\mathrm{DL}}_{\mathrm{ads,e}} \frac{\partial \Delta\Phi_{\mathrm{ads,e}}}{\partial t} = -j^{\mathrm{charge}} + j^{\mathrm{ct}}_{\mathrm{ads,e}} \tag{3.9}$$

with the potential difference at the interface $\Delta\Phi$, applied charge density j^{charge}, charge transfer reaction current j^{ct}, double layer capacitance C^{DL}, and specific effective surface area of a rough particle, $a_s^{\mathrm{eff}} = 3r_s\varepsilon_s/R_s$. Charge transfer reaction current is the sum of charge transferred at this interface. The electrical potential drop in the surface film is determined as

$$\Delta\Phi_{\mathrm{film}} = \frac{j^{\mathrm{charge}} R^{\mathrm{film}} d^{\mathrm{film}}}{a_s^{\mathrm{eff}}} \tag{3.10}$$

with specific electrical resistivity of the SEI, R^{film}, and the average film thickness d^{film}. The electrode potential $\Delta\Phi_{\mathrm{electrode}}$ is then determined as

$$\Delta\Phi_{\mathrm{electrode}} = \Delta\Phi_{\mathrm{s,film}} + \Delta\Phi_{\mathrm{ads,e}} + \Delta\Phi_{\mathrm{film}} \tag{3.11}$$

3.2.2 Atomistic scale

Whereas a macroscopic model is very limited to describe heterogeneous surface film growth processes, kMC is an adequate method [72]. Proposed mechanisms of decomposition of electrolytes at the SEI are often highly complex, where most of the reactants are not simple atoms but molecules with complex structures. The film formation is usually performed in a time period of several hours. Such a long time scale, however, can barely be realized currently by first-principles kMC simulations, which are considering every possible transition state on the atomistic scale without further simplifications. In consideration of this, the aim of applying kMC in this article is not to propose a first-principles calculation to accurately predict the formation processes, but rather to provide a methodology to introduce selected heterogeneous processes of interest into commonly used continuum-only battery models.

The kMC model is based on the solid-on-solid approach [93, 88, 118]. In contrast to a 3D model, this kMC model allows adsorption only on top of surface sites, which avoids overhangs and is a good approximation as long as difference of height between neighboring sites is not exceedingly large [118]. The kMC algorithm is based on examples provided by Burghaus [119]. A cubic lattice is used, with diagonal and horizontal diffusion allowed, while the distance between surface sites is ΔL.

The implemented processes are surface diffusion, adsorption, desorption, and reactions. Interface processes with phase change, such as electrochemical reaction, adsorption, and desorption can be anodic, cathodic, or neutral, i.e. with uncharged species. Microscopic rate Γ, i.e. transition probability, are implemented as an Arrhenius equation

$$\Gamma = A \exp\left(\frac{-E^A}{RT}\right) \tag{3.12}$$

with activation energy E^A and pre exponential factor A. For electrochemical reactions, further dependency of the applicable electrical potential $\Delta\Phi$ is considered as

$$\Gamma = A \exp\left(\frac{-E^A + \beta\Delta\Phi\mathrm{F}}{RT}\right). \tag{3.13}$$

Here shown at the example of an anodic reaction. Diagonal and horizontal surface diffusion pre-exponential factors are determined based on the surface diffusion coefficient D, with pre-exponential factor being $A^{\text{diff}} = w^h = D/2(\Delta L)^2$ for the horizontal diffusion and $A^{\text{diff}} = w^d = D/4(\Delta L)^2$ for diagonal diffusion [93]. The activation energy for diffusion or desorption steps depends on the bonds to nearest neighbor sites and can be calculated ads $\sum_k n^k E^A_{\text{bond}(k)}$, where n^k is the number and $E^A_{\text{bond}(k)}$ is the bonding energy to a neighbor of type k [88]. The adsorption rate depends on the species activity in the electrolyte $a^{k(\mathrm{e})}$. Some of the reactions involve electrons which are present at the electrode surface; those electrochemical reactions are driven by the potential difference between electrode and the adsorption site. Furthermore, electrons need to pass the surface film and thus the reaction rates decrease with increasing film thickness. The electron leakage process is not fully understood. Even so, some suggestions are given in the literature [84, 83, 120, 82]. In this work, we assume that an electron needs to overcome an additional activation energy E^A_{film}, which depends on the specific activation energy $\hat{E}^A_{\text{film}}$ and the local film thickness $\hat{d}^{\text{film}}$ between local surface site and the electrode as

$$E^A_{\text{film}} = \hat{d}^{\text{film}} \hat{E}^A_{\text{film}}. \tag{3.14}$$

As such, the reaction rate exponentially decreases with increasing surface film thickness. This dependency is often assumed in macroscopic models [28]. The specific activation energy $\hat{E}^A_{\text{film}}$ is thereby a parameter that is chosen to achieve a reasonable film thickness. Reverse rates of processes in this model as well as in the macroscopic model are determined based on the standard chemical potentials of the species, by calculating the change of Gibbs free energy $\Delta G^0 = \sum \mu^0_{\text{products}} - \sum \mu^0_{\text{educts}}$ [117], which can then be further used to determine the backward reaction rates from

$$\frac{k^f}{k^b} = \exp\left(\frac{-\Delta G^0}{RT}\right) \tag{3.15}$$

As mentioned previously, the kMC simulations are used to introduce heterogeneous surface processes into a continuum model. As such, only particular species of interest are considered explicitly by the kMC method while others are considered via the continuum model. The probability of a species on a surface site is approximated by a conditional expectation based on the deterministic solution of the macroscopic equations. Details are shown in chapter 2.

3.2.3 Multiscale coupling

The main methodology, algorithms, and considerations of the multiscale coupling have been already provided in chapter 2. This chapter discusses only particular aspects which are important for interpretation of simulation results and setup for the multiscale simulation of this particular problem.

The macroscopic model and the atomistic model are dynamically coupled by using MPA1, as given in chapter 2. An important aspect of a multiparadigm method are the data which are actually exchanged. This multiscale coupling allows the coupling of detailed heterogeneous surface film growth mechanisms with the continuum model. The continuum model thereby applies mass balance equations, which contains net reaction rates based on kMC rates Q^{kMC}, which are determined as

$$Q^{\mathrm{kMC}} = \frac{\Delta n_i}{\Delta t_i A^{\mathrm{kMC}}} \tag{3.16}$$

where Δn_i is the total moles converted during the simulation period Δt_i at the surface area of the kMC model A^{kMC}. To improve the stability of the numerical solution, the rates Q of the continuum model were not directly set to the rates of the kMC model, but instead the main impact of the potential difference $\Delta\Phi$ and the average film thickness d^{film} was considered in the macroscopic model by

$$Q = \lambda^{\mathrm{kMC}} \exp\left(\frac{0.5\Delta\Phi_{\mathrm{s,film}}\mathrm{F} - d^{\mathrm{film}}\hat{E}^{A}_{\mathrm{film}}}{\mathrm{R}T}\right) \tag{3.17}$$

which is then corrected in every iteration by an adjustable parameter λ^{kMC} to fit rates of the kMC simulation. Further, the time step of a multiscale iteration is adapted based on the kMC steps of the previous iteration. A low number of kMC steps in the previous iteration step increases the length of the next time step. Dependent on the particular simulation, the robustness of the multiscale model can be sensitive on coupling strategy, such as iteration time step and data filter type [14], which has been discussed in detail in chapter 2. The whole code including macroscopic and atomistic model is implemented in MATLAB. Partial differential equations are solved by the finite volume method and adaptive ODE solvers, i.e. ode15s.

3.3 Example problem: Film growth in ethylene carbonate based electrolyte

To demonstrate the capability of the multiscale model to analyze layer growth and its interaction with the macroscopic states, simulations are performed for an example problem. The decomposition of EC at a graphite electrode surface is chosen. The reaction mechanism is built on literature findings. As the mechanism is still under discussion, it should be seen as an example that can be adjusted later, once the full mechanism is known. The mechanism as implemented in the macroscopic and atomistic model is illustrated in Figure 3.2. Details about the concept, as well as assumed surface film components and reactions, are given below.

3.3.1 General simulation concept

The implemented reaction mechanism is shown in Table 3.1 along with the respectively applied simulation methods and kinetic parameters. Properties of the species involved

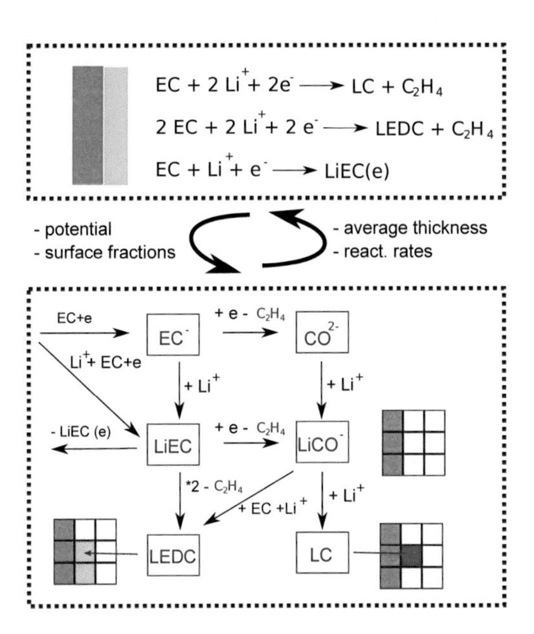

Figure 3.2: Illustration of the assumed reaction mechanism in the continuum model (top) and the kMC model (bottom), including information exchanged to synchronize those reactions in the multiscale simulations. Reprinted from publication [106].

in the kMC and continuum model are given in the appendix A in Table A.2. Properties of the species involved in the kMC and continuum model are given in the appendix A in Tables A.3 and A.4, respectively. In the continuum model, species can be located in different phases, which is indicated in brackets. In the kMC model, the species Li^+, EC, PF_6^-, and e^- are not considered explicitly, but occupy free surface sites with a certain probability as explained in chapter 2. The intermediate species are initially generated on the surface via reactions of those species, i.e., reactions 1 and 2. When generated, intermediate species may diffuse on the surface, while probability of a diffusion step depends on the binding to its nearest solid neighbors. An essential prerequisite to obtain experimentally observed nano-structures is that reactions need to be favored differently, depending on the local surface configuration. A physical reason for these differences may be a selective binding of species to certain solid components. It is assumed that intermediate $LiC_3H_4O_3$ and $LiCO_3^-$ have only low binding energies to the solid LC and LEDC, respectively, which leads to the formation of such heterogeneous structures of the film. The concentration of intermediate species in the electrolyte is negligible compared to salt and solvent concentration, and is set to zero.

As can be seen in Figure 3.2, net reactions considered in the continuum model are direct pathways from the reactants EC, Li^+, and e^- to the products C_2H_4, lithium carbonate (LC), lithium ethylene dicarbonate (LEDC), and the desorbed intermediate species LiEC. To specify flow rates Q, the parameter λ^{kMC} as well as average film thickness is provided by the kMC model. Continuous change of system states is considered via modeling surface fractions θ of the reactants Li^+, PF_6^-, and EC as well as electric potential $\Delta\Phi_{s,ads}$ by the continuum model.

3.3.2 Solid components within the surface film

The assumed solid products are LC and LEDC, which have been reported as being the main SEI film components [82, 76, 8, 80, 112, 121]. LC is reported to be the most important component for SEI functionality [84, 122] and formed with EC-based electrolytes [112]. A two-layer structure of the SEI is commonly reported [84, 123], with a dense inner film and a porous outer film [123]. In this work, a kMC model is used for the growth mechanism of the dense inner film. The outer porous film is considered by assuming all desorbed intermediates will react in the electrolyte and form solid components. Those solids will reduce the porosity of the electrode, which will impact macroscopic transport processes in the porous electrode during operation of the battery. During the formation process, only very low currents are applied and thus spacial distribution of solvent and lithium due to transport processes in the porous electrode are neglected in this chapter. However in principle it can be considered, as will be shown in chapter 4.

The size of the lattice elements in the kMC model is determined based on LC, which is a good lithium-ion conductor and therefore the most relevant surface species [82]. LC forms a crystal structure with a monoclinic lattice system with $a = 8.370$ Å, $b = 4.929$ Å, $c = 5.870$ Å, and $\beta = 117.1°$ [82]. A cubic lattice would achieve the same volume with lateral length of $a = 5.996 \approx 6$ Å. Thus the distance between lattice sites is defined to be $\Delta L = 6 \times 10^{-10}$ m. The organic solid LEDC is reported to be the most common product observed in the SEI experimentally [80, 124]. This species consists of two CH_2OCO_2Li groups, each of which are approximately as large as LC. Therefore, LEDC is assumed to fill two surface sites and has the same height as LC. This assumption enables the representation of the complex film structure in a simple cubic lattice.

3.3.3 Reaction mechanism

The heterogeneous reaction mechanism applied here is abstracted from the literature. EC ($C_3H_4O_3$) is reported to break down by consuming an electron through the breakage of one of the C–O bonds next to the C_2H_4 group to form $C_3H_4O_3^-$ via reaction 1 [76, 77, 80]. An alternative transition state for the bond breakage is suggested by [80], where the energy barrier is lower in the presence of lithium, leading to the intermediate product $LiC_3H_4O_3$ via reaction 2. The reduced species $C_3H_4O_3^-$ and $LiC_3H_4O_3^-$ can then react with a further electron to form CO_3^{2-} via reaction 3 and $LiCO_3^-$ via reaction 4, respectively. In both reactions, the gaseous species C_2H_4 is produced. For reaction 3, a transition state with a very high energy barrier is suggested [80]. Most literature agrees that CO_3^{2-} quickly reacts further with lithium to form first $LiCO_3^-$ via reaction 6 and finally the solid LC (Li_2CO_3) via reaction 9 [76, 114, 77, 80, 125, 121]. Except for reaction 7, those reactions do not involve more than one of the rare intermediate species, therefore it can be expected that species $C_3H_4O_3^-$, CO_3^{2-}, and $LiCO_3^-$ have very short lifetimes, which indicates that desorption for those species does not need to be considered. LEDC ($(CH_2OCO_2Li)_2$) is produced either through reactions 7 or 8. In reaction 7, LEDC is formed by the reaction of two $LiC_3H_4O_3$ and in reaction 8 with Li, EC, and $LiCO_3^-$. Gibbs free energies of the reactions are partly provided by Wang et al. [80]. Standard state chemical potentials and remaining reaction energies are chosen based on the data where applicable and otherwise chosen in the adequate order of magnitude. A quantitative computational determination of all parameters for this reaction mechanism via DFT simulations is yet not possible [91] and remains the subject of future studies. Additional parameters as applied in those simulations within this chapter are provided in the appendix A in Table A.5.

As the parameters and mechanisms are not fully available in literature, the results should be considered as a first qualitative insight into the formation process and an assessment of the multiscale nature of SEI growth, and as a demonstration of the multiscale model as a way to evaluate hypothesized mechanisms.

3.4 Discussion of the simulation results

The multiscale model is used to simulate the processes during the first charge for two cases that allow the assessment of the effect of electrodes with different macroscopic properties: particle size $R_1 = 3$ µm, i.e. fine electrode, and $R_2 = 10$ µm, i.e. coarse electrode. In both cases, formation is performed with typical low constant charge rate of 0.1 C. Detailed results for this process are shown and the effect of macroscopic properties on atomistic surface reaction and film growth and vice versa are discussed.

The potential of the graphite electrode during the first charging process is shown

Table 3.1: Reaction processes including reaction rate constant k, Gibbs free energy ΔG^0, and activation energy E^A. Reprinted from publication [106].

Number	Reactions	method
1	$C_3H_4O_3 + e^- \rightleftharpoons C_3H_4O_3^-$	kMC
2	$C_3H_4O_3 + e^- + Li^+ \rightleftharpoons LiC_3H_4O_3$	kMC
3	$C_3H_4O_3{}^- + e^- \rightleftharpoons CO_3^{2-} + C_2H_4$	kMC
4	$LiC_3H_4O_3 + e^- \rightleftharpoons LiCO_3^- + C_2H_4$	kMC
5	$C_3H_4O_3^- + Li^+ \rightleftharpoons LiC_3H_4O_3$	kMC
6	$CO_3^{2-} + Li^+ \rightleftharpoons LiCO_3^-$	kMC
7	$2\ LiC_3H_4O_3 \rightarrow (CH_2OCO_2Li)_2 + C_2H_4$	kMC
8	$LiCO_3^- + Li^+ + C_3H_4O_3 \rightarrow (CH_2OCO_2Li)_2$	kMC
9	$LiCO_3^- + Li^+ \rightarrow Li_2CO_3$	kMC
10	$Li(s) \rightleftharpoons V(s) + Li^+(s) + e^-(s)$	continuum
11	$Li^+(s) \rightleftharpoons Li^+(ads)$	continuum
12	$Li^+(ads) \rightleftharpoons Li^+(e)$	continuum
13	$PF_6^-(ads) \rightleftharpoons PF_6^-(e)$	continuum
14	$C_3H_4O_3(ads) \rightleftharpoons C_3H_4O_3(e)$	continuum
15	$2\ C_3H_4O_3(ads) + 2\ Li^+(ads) + 2\ e^-(ads) \rightarrow (CH_2OCO_2Li)_2 + C_2H_4(e)$	continuum
16	$C_3H_4O_3(ads) + 2\ Li^+(ads) + 2\ e^-(ads) \rightarrow Li_2CO_3 + C_2H_4(e)$	continuum
17	$C_3H_4O_3(ads) + Li^+(ads) + e^-(ads) \rightarrow LiC_3H_4O_3(e)$	continuum

in Figure 3.3 and reveals the typical features of the formation process, such as the formation plateau at $\sim$ 0.6 V, as well as the capacity needed for the first charge including the formation process. The slope of the potential is dependent on the particle size. The main differences between the curves can be explained as follows. Smaller particles provide larger surface area so more solid electrolyte is formed, which results in a more distinct plateau at the beginning of charge and a higher capacity, i.e. lithium consumption, during the first charge. Furthermore, the formation process in the larger particle starts at lower potentials, which is due to higher overpotentials of the electron leakage process because of smaller surface area.

While it is possible to provide sound explanations for almost all of the main features that can be observed in the potential curve, it is more difficult to give explanations for the characteristics of the film growth, which is shown in Figure 3.4. This figure shows the average film thickness of both electrodes during the charging process for the dense inner film, which limits the passing of electrons. The figure does not include the intermediates desorbed to the electrolyte. Those species are instead expected to further react in the electrolyte phase and form a second porous film. This part of

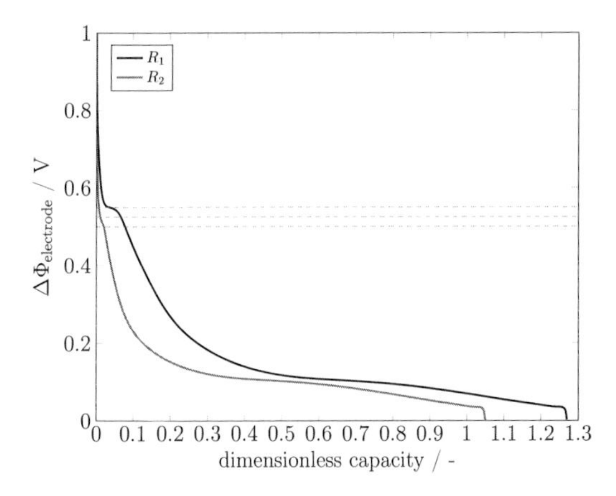

Figure 3.3: Electrode potential $\Delta\Phi_{\text{electrode}}$ during the first charge of the formation process for electrodes with particle radius $R_1 = 3 \times 10^{-6}$ m and $R_2 = 10 \times 10^{-6}$ m. Dashed gray lines indicates the potentials 0.55 V, 0.525 V, and 0.5 V from top to bottom. Reprinted from publication [106].

the SEI would affect diffusion processes in the porous electrode, but, due to very low charging rates, this effect is not further considered in this simulation. For both electrodes, the evolution of the film growth is qualitatively similar. While the growth is steep at the beginning of charging, the growth slows down at the end of charging. It has been stated that the film grows approximately with the square root of time [33, 110], which cannot be seen for the simulation results given here, but may be the long time trend for a life cycle test. In case of the electrode with large particles, the film grows faster compared to the electrode with small particles. This observation is again related to the smaller surface area provided by larger particles, which leads to much higher local surface current density and thus faster formation of the surface film. This slope of film thickness is similar to the inverse of the potential for both electrode, which indicates the strong impact of electrode potential on the film growth. Further, the thickness of surface film for both electrodes is almost the same at the end of discharge (0 V), even though the potential is reached at considerably different discharge capacities. The particle size impacts growth rate much more than final thickness. In addition, it can be seen that the film is not growing at the very beginning of the charging, while the electrode potential in Figure 3.3 during this period stagnates, i.e. shows the typical formation plateau. The potential starts to drop as soon as a stable dense inner film is formed, which limits further electrochemical side reactions. Further detailed explanation is given later in the discussion.

The heterogeneous film structure and its growth are shown in Figure 3.5 for both

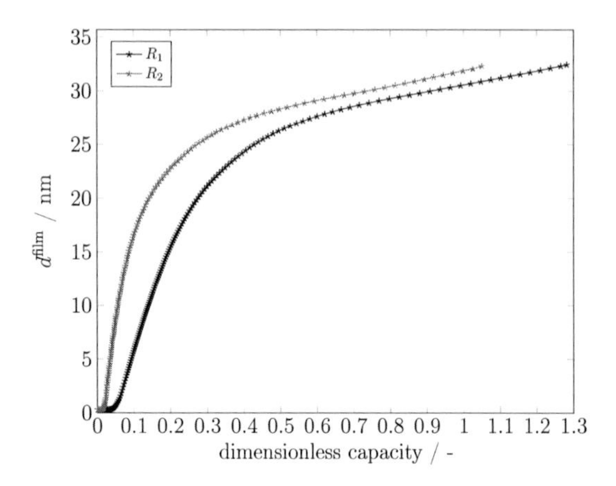

Figure 3.4: Thickness of the dense inner film during the first charge of the formation process for electrodes with particle radius $R_1 = 3 \times 10^{-6}$ m and $R_2 = 10 \times 10^{-6}$. Reprinted from publication [106].

electrodes at the same potentials. In both systems, LC and LEDC solids are the main products, while a nano-structuring of the solids can be observed. The structuring is a direct consequence of the different binding energies of the intermediates. The surface film shows some extent of roughness, but there are no areas with local heights that considerably deviate from the average film thickness. The reason for this observation is that resistance of the electrochemical side reactions is evaluated locally. Therefore, electrochemical processes have a higher probability to occur at regions close to the anode surface, which leads to an intrinsic adjustment of the local film thickness. If only some of the solid components allow electron leakage through the surface film, locally diverging growth rates can lead to less planar surfaces. However, those effects were not considered in this study.

In the following, the structure and the growth process for coarse and fine electrodes, which is shown in Figure 3.5 is discussed. The formation of the solid film starts earlier with the fine electrode, which can be seen in Figures 3.5ABC and 3.5EFG and also in Figure 3.4. At the end of charge, i.e. 0 V (Figures 3.5DH), it again can be seen that final film thickness is almost independent of particle size. For both electrodes at the beginning of the formation process, several islands are formed at the anode surface, which then further grow until the whole surface is covered. It can be seen that LC and LEDC components are formed close to each other, while LEDC is located in the center and LC is at the edges of the islands. At the beginning of the formation, i.e. Figures 3.5AF, LC can be also observed detached from LEDC, while LEDC seems to grow only on top or next to LC components. This observation can be explained by

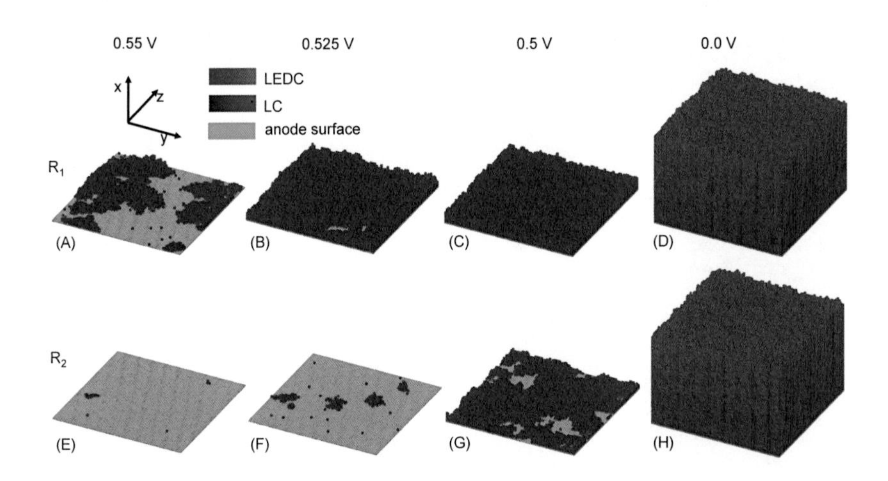

Figure 3.5: KMC configurations of the dense surface film for electrode with particle radius $R_1 = 3 \times 10^{-6}$ m (A–D) and $R_2 = 10 \times 10^{-6}$ m (E–H) at different electrode potentials with 0.55 V (A,E), 0.525 V (B,F), 0.5 V (C,G), and 0.0 V (D,H) during the first charge of the formation process. Reprinted from publication [106].

reaction 7, where LEDC is formed by two $LiC_3H_4O_3$ components on the surface. The probability for two of those species being next to each other is higher at regions where they bind to several solids than on the initially plain surface where they only bind to one solid component. Comparing the final structure of both electrodes in Figures 3.5DH, it can be seen that the structures differ. However, these differences correlate to the stochastic nature of the process rather than to a significant impact of particle size. This may be due to low C-rates of the formation process and could be more distinct for faster formation procedures or during operation of the cell. No considerable impact of particle size on SEI structure could be observed in this simulation study.

Even if the final SEI structure is not significantly influenced by the particle size, a significant difference in the overall amount of the produced SEI components vs. total electrode volume can be observed in Figure 3.6. Comparing the volume fraction of electrode with small particles R_1 and large particles R_2 shows that about 5 times more SEI components have been formed for electrodes with small particles, which correlates also to the higher charge capacity in Figure 3.3, which is about 5% and 25% of the theoretical capacity for small and large particles, respectively. In both electrodes, the amount of porous SEI formed is about 2 times that of the dense SEI, i.e. LC and LEDC. For the cell with small particles, the fraction of solid components produced is very high and thus would considerably affect electrode performance during operation

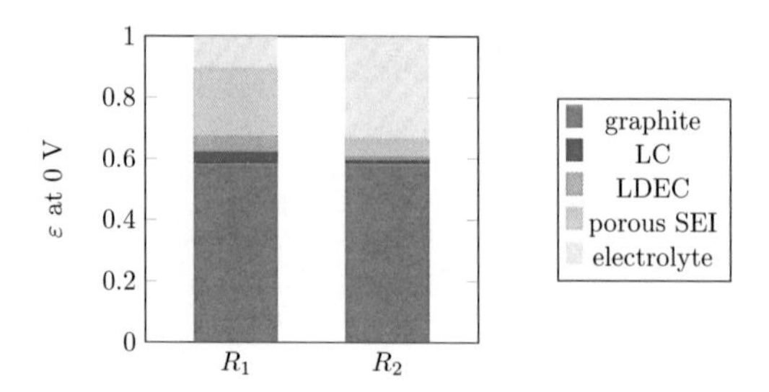

Figure 3.6: Volume fraction of electrode component at the end of charge, i.e. 0.0 V (A) for electrodes with particle radius $R_1 = 3 \times 10^{-6}$ m and $R_2 = 10 \times 10^{-6}$ m (C). Reprinted from publication [106].

with higher C-rates.

The process rates for LEDC, LC, and desorption (des) at the surface are shown in Figure 3.7 for the kMC (marks) and the macroscopic model (lines). The rates are in good agreement, which is an indicator for good quality of the coupling. Comparing the quantity of the rates, the local reaction rates are seen to be considerably higher for the larger particles (B) compared to smaller particles (A), which is mainly due to the difference in surface area. For both electrodes, first the rate for intermediates desorbing increases rapidly, which is then followed with a delay by the actual production of the LEDC and LC component forming the dense inner film. For both electrodes, the rate of LEDC is higher compared to the rate of LC. As soon as the film covers the whole surface, the desorption rate decreases, which can be explained by the change of the surface properties from a flat ground to a rough film. A rough structure provides more binding energy for the intermediate components, which increases the probability of the following reactions. With ongoing film growth, reaction and desorption rates decrease due to a higher activation energy for the electron leakage as applied for electrochemical side reactions. This decrease of film growth reactions is faster and steeper for large particles (R_2). Further, for smaller particles this decrease is considerably slower in the beginning, which cannot be observed for the large particles. The magnitude of all three side reactions is similar in this simulation. For the LC rate in the beginning of charge, a steep overshooting can be observed for large particles, which is considerably more flattened for small particles. Even so, the general trend is similar for both electrodes, and it can be seen that the macroscopic properties considerably impacts the local reaction rates at the surface during the formation process.

To further understand the presented results and multiscale coupling, the data exchanged between the models are analyzed. As explained previously, the reaction rates in the macroscopic model are not set directly to those determined by the kMC model.

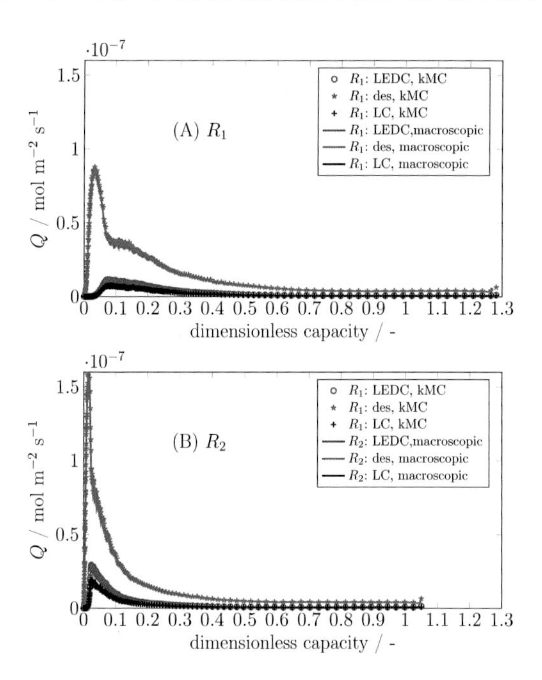

Figure 3.7: Reaction rates for LEDC, LC, and desorption (des) at the surface for electrode with particle radius $R_1 = 3 \times 10^{-6}$ m (A) and $R_2 = 10 \times 10^{-6}$ m (B) including the results of the macroscopic (lines) and the kMC simulation (marks). Reprinted from publication [106].

Instead reaction rates are defined in equation 3.17 and adjusted to fit the reaction rates of the kMC model. The only adjustable parameter in this equation is λ^{kMC} for reaction 15-17, which is determined after each multiscale iteration and thus represents the information actually passed from the atomistic to the macroscopic model. Figure 3.8 shows the parameter λ^{kMC} for reactions 15–17. For all three processes, fluctuations of this parameter can be seen. Fluctuation of the kMC output indicates the uncertainty of a process, which is strongly related to its frequency. As long as a process occurs frequently, statistics are reliable and fluctuation is low. The fluctuation may be decreased by increasing surface area A^{kMC}, but with an increase in computational time. As can be seen, all transferred parameters change during the formation process. The parameter for the rate equation of LEDC is shown in Figure 3.8A. The value rapidly increases at the beginning of the formation process and then only slightly changes, while changes are higher with higher potential gradients. In Figure 3.8B, the parameter for rate equation of LC reaction is shown, which is qualitatively

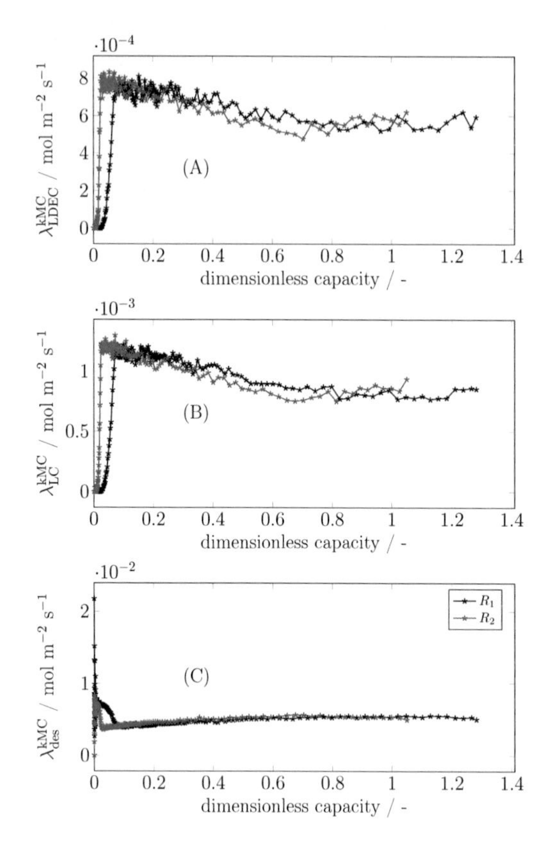

Figure 3.8: KMC output for calculation of reaction rates of LEDC (A), LC (B), and desorbed intermediates (C) for electrodes with small particles (black) and large particles (blue). Reprinted from publication [106].

comparable to that of LEDC. In Figure 3.8C, the parameter for the rate equation of the desorption of $LiC_3H_4O_3$ is seen to rapidly decrease as soon as a stable dense inner film is formed. This decrease starts earlier for the larger particles. Further, the value is seen to change during the whole charging process. In general, a change of these parameters denotes a correction of assumed dependencies of equation 3.17. A constant value would denote that there is no multiscale interaction between the atomistic and the macroscopic model. For the mechanisms shown, the assumed dependency of the reaction rate from electrode potential and average film thickness seems to be good at the later state of the charge process, but not for the first part of the charging where the structure of the surface film, e.g. roughness, undergoes considerable changes.

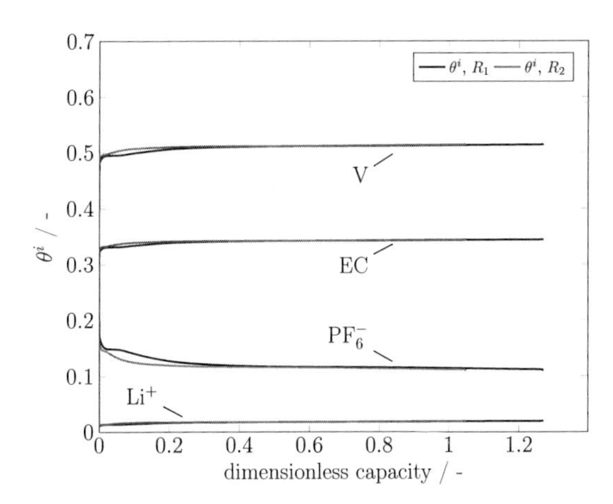

Figure 3.9: Surface fraction of major species at the adsorption film for electrodes with small particles (black) and large particles (blue). The states are calculated by the macroscopic model and fed to the atomistic model. Reprinted from publication [106].

The surface fractions provided by the macroscopic model are shown in Figure 3.9. The system states passed from the macroscopic to the atomistic model are the surface fractions of the species (Li^+, PF_6^-, EC, and vacancies V) as well as the electrode potential gradient $\Delta\Phi_{s,ads}$ between the electrode and the adsorption site. The surface fractions of these species are very similar for both structures, because they mainly depend on the concentrations in the electrolyte, which is constant in this simulation. The most influential exchanged state is the electrode potential, which is shown in Figure 3.3. The electrode potential directly impacts all of the electron-involving reactions on the surface, and is the variable that triggers and determines the multiscale interaction between the macroscopic and the atomistic model. Through the electrode potential, the macroscopic conditions directly impact the evolution of the atomistic system, i.e. surface film growth and structure. The structure of the surface film determines the rate constants, i.e. electron consumption rates, as used in the macroscopic model and thus impacts the evolution of macroscopic properties such as electrical potential or average film resistance. Both lead to a continuous multiscale interaction between the macroscopic and the atomistic model.

3.5 Concluding remarks

In this chapter, multiparadigm modeling as shown in chapter 2 has been applied to perform multiscale simulations for analysis of SEI formation in lithium-ion batteries. The model dynamically couples a macroscopic electrode model including electron leakage limitation through a surface film with a kinetic Monte Carlo model for the simulation of atomistic degradation processes on the surface as well as heterogeneous film growth. Simulations are carried out on the example that assumes EC as solvent and $LiPF_6$ as salt. Two different particle sizes were used to reveal the effect of different macroscopic system properties.

The presented multiscale modeling approach enables the coupling of heterogeneous growth mechanisms with continuum models. The macroscopic properties directly impact the atomistic processes, where lateral interactions lead to formation of nanostructures on the surface that result in different macroscopic side reaction rates at the surface and thus affect the macroscopic system. Furthermore, the results provide some suggestions on functionality of surface film growth. The film growth is shown to be considerably faster with larger particles, while the final thickness of the electrode correlates to the electrode potential and thus is similar for both electrodes.

The results indicated that the size of particles impact the SEI formation process. In an actual electrode there are many particles of different size, i.e. particle size distribution. It can be expected that the actual particle size distribution has an impact on local states and thus local degradation mechanisms such as the SEI formation process. A detailed analysis of the impact of such an macroscipic heterogeneity on local conditions, performance and degradation is provided in chapter 5.

With this chapter it has been shown how hypothetical microscopic reaction mechanisms can be tested. The method enables new possibilities towards understanding, predicting, and optimizing SEI formation by adapting macroscopic properties, electrolyte composition, or charging strategies. Nevertheless, in this chapter only a simplified model of a porous electrode, i.e. SP model, was applied. It is well known that there is a spacial distribution of concentrations and potentials along the thickness of the electrode, which is considered in the most commonly applied continuum model for lithium-ion batteries, i.e. the pseudo two dimensional model. In particular at faster formation procedures it can be expected that film growth and structure will depend on position within an electrode. As structure and thickness determines the film resistance the flow of charge carries will be influenced. Further, the parameters of the model need to be obtained in order to analyze particular electrodes. Both issues are addressed in detail in chapter 4 to achieve a development towards analysis of multiscale effects in actual technical cells.

Chapter 4

Multiscale Analysis of Film Formation [5]

4.1 Introduction

In chapters 2 and 3 the multiparadigm methodology was developed and applied at an example problem of surface film formation in a lithium-ion battery electrode. With this, the technical basis for a multiscale analysis of film formation problems in batteries is given. Thus this chapter applies the developed methods for analysis of an actual technical cell. Parameters are identified from an engineering perspective, i.e. based on electrochemical measurements. In chapter 3, it has been shown that the method in general is feasible to describe and analyze multiscale problems, while in this chapter evidence for multiscale effects, as described in chapter 1, is provided using the parameterized model.

SEI microstructure and compostion has a significant impact on cell performance [126]. It depends on graphite structure, electrolyte composition, temperature and electrochemical conditions, e.g. cut-off-potential and current densities [109]. As will be shown with this chapter, there is a need for a detailed understanding of the growth process and the multiscale interaction in order to optimize the cell production, e.g. the formation protocols. In the following, gerneral aspects of the multiscale interaction are summerized based on literature findings. Märkle et al. [127] showed that the film morphology can be controlled by the current in the first cycle and has considerable impact on the cycling stability of the cell. The impact of temperature and upper cut-voltage voltage has been shown by German et al. [128]. Further, Antonopoulos et al. [129] carried out formation experiments with constant potential and demonstrated a significant impact on the chemical composition of the SEI as well as the performance of the cell after formation. It has been demonstrated that the SEI can have non-homogeneous structure, consisting of several SEI microphases

[5] Part of this chapter has been published in (Röder et al., Batteries and Supercaps, 2:248–265, 2019 [1]) and is reproduced with permission from Wiley-VCH Verlag GmbH & Co. KGaA

[115, 130]. Chattopadhyay et al. [108] experimentally observed LiF crystals within the SEI and suggest that their orientation may modify Li^+ diffusion and thus impacts the performance of the battery. The formation of clusters have been also observed by molecular dynamic (MD) simulations [131]. As degradation and cell failure is often triggered by heterogeneity on nano- and mesoscale [132, 133], it is essential to study and understand its origin in order to achieve optimal performance of the battery.

This chapter provides detailed descriptions of model equations needed fur multiscale analysis at high C-rates. For this, the most commonly applied battery model, i.e. the P2D model, is extended by an heterogeneous surface film growth model using kinetic Monte Carlo simulations analogously to chapter 3. Lumped hypothetical mechanism on atomistic scale are introduced, while rate expressions are derived from macroscopic reversible thermodynamic and kinetic expressions. This enables identification of key parameters of the atomistic model and good agreement to the electrochemical experiments. The whole procedure is outlined using experimental data of a lithium-ion battery.

4.2 Multiscale analysis

4.2.1 Experimental

For experimental characterization, electrodes were produced in the Battery LabFactory Braunschweig. The cell was assembled and characterized in a three electrode setup using the PAT-cell format of the EL-cell company. As active materials, commercial $Li(Ni_{1/3}Co_{1/3}Mn_{1/3})O_2$ and graphite (SMG) were used as the positive electrode (cathode), and the negative electrode (anode), respectively. Electrodes were produced using carbon black, graphite, binder (PVDF), and the active material in the respective ratio 4:4:2:90, for the cathode and 5:2:2:91, for the anode. The ratio is thereby given in weight-%. Cells were filled with electrolyte using ethylene carbonate (EC), dimethyl carbonate (DMC), and ethyl methyl carbonate (EMC) solvents in the ratio 1:1:1 with 1 M $LiPF_6$ including 2% vinyl carbonate (VC) and 3% cyclohexyl benzene (CHB).

The cell was characterized electrochemically in two regions, during the first formation cycles, where significant contribution of side reactions occur, and after the initial film formation, where no significant contributions of side reactions is expected. The experimental data for formation, open circuit potential and C-rate tests, are used for parameter identification of the electrochemical model. Formation was performed with 2.15 A m^{-2}. OCP was measured by incrementally discharging in steps of 0.05 V. The C-rate test was performed for cell discharge with 4.2, 21.5, and 64.5 A m^{-2} between 4.2 and 2.9 V.

4.2.2 Concept for model based analysis

In this chapter, multiscale interaction is studied for the lithium-ion battery given above by doing a model based analysis of electrochemical measurements. In Figure 4.1 (A), a lithium-ion battery is illustrated on multiple scales, i.e. electrode scale, particle scale, mesoscale, and atomistic scale. The multiscale nature and interaction between operational and side reaction processes has been outlined in chapter 1. With this in mind, the concept and the scope of the presented model based multiscale analysis is given.

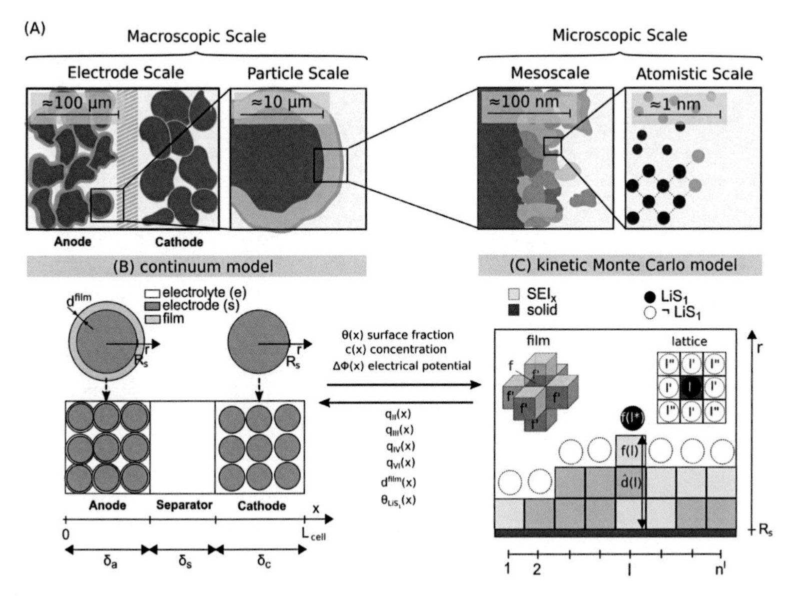

Figure 4.1: Typical scales within lithium-ion battery cell (A), continuum model (B), and kMC model (C). Reproduced from [1] with kind permission from Wiley-VCH Verlag GmbH & Co. KGaA.

Usually, side reactions can barely be detected electrochemically in a single charge/discharge cycle. Because measurements are dominated by the operational processes, their impact can finally be observed as capacity or power fade in ageing experiments. However, at the first charge, i.e. during SEI formation, side reactions are one of the main reaction processes occurring at the anode. This is used in this chapter by applying a two step parameterisation procedure of the model, which is described in the following. First, the operational processes, intercalation de-intercalation, ion and electron transport in solid and electrolyte are identified using measurements of the open circuit potential and C-rate tests. As these measurements take place after film formation, it is assumed that there is no significant contribution of side reactions to these measurements. In a second step, the parameters of the side reaction process are identified using the first charge cycle of the battery. As parameters of the normal operation after formation have been identified already, a separation of the influence of the side reaction process and thus a separate and more reliable parameter identification is enabled.

All models described in the following sections as well as model coupling is implemented in MATLAB. Thereby, differential equations are solved using adaptive solver ode15s. KMC simulations are always carried out with 9 parallel instances on a 30×30 lattice.

4.3 Macroscopic scale

4.3.1 Continuum model

The continuum model is illustrated in Figure 4.1 (B). The model is based on the P2D model as first proposed by Doyle et al. [21]. Additionally, charge balances at electrochemical double layers and a thin film at the particle surface are included based on Legrand et al. [134] and Colcasure et al. [30], respectively.

The partial and ordinary differential equations are given in Table 4.1 and cover from top to bottom: mass balance of lithium in solid including diffusion of lithium in the spherical particles, species balance of salt within the electrolyte including diffusion and reaction, balance of dissolved uncharged species in the electrolyte, charge transport in the electrode including diffusion and reaction, charge balance in the electrolyte, charge balance at the electrochemical double layer at the solid/surface film (s/film) interface, charge balance at the electrochemical double layer at the adsorption site/-electrolyte (ads/e) interface, and the balance of species on the adsorption layer. The time dependent porosity has been taken out of the time derivative, which is a feasible approximation in consideration of slowly changing porosity. The adsorption process of lithium ions from the liquid electrolyte to the surface of the SEI is treated analogously to an electrochemical reaction as suggested by Schleutker et al. [135]. Macroscopic and microscopic processes considered at the interfaces are summarized in Table 4.2. Process I and XI correspond to the normal operation of the battery. Decomposition mechanisms, i.e. Equation 4.42-4.45, are included with process II, III, IV, V, and XII. Further, sorption processes for solvents and intermediates are included with process VIII, IX, X. Surface diffusion processes are given with XII and XIII. In the continuum model, only the macroscopic processes are considered. Processes which are given as microscopic and as macroscopic processes, i.e process II, III, IV, and VI, are covered by the kMC model, while process fluxes q are synchronized with the continuum model as illustrated in Figure 4.1. Considered species can be found in the appendix A in Table A.6, while species covered by the continuum method are indicated accordingly. In the following, details about equations and parameter identification is provided.

Table 4.1: Set of partial and ordinary differential equations of the applied P2D continuum battery model with m being the desolved species int he electrolyte. Reproduced from publication [1] with kind permission from Wiley-VCH Verlag GmbH & Co. KGaA.

Equation	Boundary Condition
$\frac{\partial c_s^{\mathrm{Li}}(t,r)}{\partial t} = \frac{1}{r^2}\frac{\partial}{\partial r}\left(D_s^{\mathrm{Li}} r^2 \frac{\partial c_s^{\mathrm{Li}}(r)}{\partial r}\right)$	$\begin{cases} -D_s^{\mathrm{Li}}\frac{\partial c_s^{\mathrm{Li}}(0)}{\partial r} = 0 \\ -D_s^{\mathrm{Li}}\frac{\partial c_s^{\mathrm{Li}}(R_s)}{\partial r} = -r_s q_{\mathrm{I}}(x) \end{cases}$
$\varepsilon_e(x)\frac{\partial c_e^{\mathrm{salt}}(x)}{\partial t} = \frac{\partial}{\partial x}\left(D_{e,\mathrm{eff}}^{\mathrm{salt}}(x)\frac{\partial c_e^{\mathrm{salt}}(x)}{\partial x}\right) + (1-t_p)j_{\mathrm{ads,e}}^{\mathrm{ct}}(x)$	$\begin{cases} \frac{\partial c_e^{\mathrm{salt}}(0)}{\partial x} = 0 \\ \frac{\partial c_e^{\mathrm{salt}}(L_{\mathrm{cell}})}{\partial x} = 0 \end{cases}$
$\varepsilon_e(x)\frac{\partial c_e^{m}(x)}{\partial t} = \frac{\partial}{\partial x}\left(D_{e,\mathrm{eff}}^{k}(x)\frac{\partial c_e^{m}(x)}{\partial x}\right) + a_s r_s q_{\mathrm{ads,e}}^{m}(x)$	$\begin{cases} \frac{\partial c_e^{m}(0)}{\partial x} = 0 \\ \frac{\partial c_e^{m}(L_{\mathrm{cell}})}{\partial x} = 0 \end{cases}$
$0 = \frac{\partial}{\partial x}\left(-\sigma_{s,\mathrm{eff}}\frac{\partial \Phi_s(x)}{\partial x}\right) + j(x)$	$\begin{cases} -\sigma_{s,\mathrm{eff}}\frac{\partial \Phi_s(0)}{\partial x} = I_{\mathrm{cell}} \\ -\sigma_{s,\mathrm{eff}}\frac{\partial \Phi_s(\delta_a)}{\partial x} = 0 \\ -\sigma_{s,\mathrm{eff}}\frac{\partial \Phi_s(L_{\mathrm{cell}}-\delta_c)}{\partial x} = 0 \\ -\sigma_{s,\mathrm{eff}}\frac{\partial \Phi_s(L_{\mathrm{cell}})}{\partial x} = -I_{\mathrm{cell}} \end{cases}$
$0 = \frac{\partial}{\partial x}\left(-\sigma_{e,\mathrm{eff}}(x)\frac{\partial \Phi_e(x)}{\partial x} - \sigma_{\mathrm{De,eff}}(x)\frac{\partial ln(c_e^{\mathrm{salt}}(x))}{\partial x}\right) - j(x)$	$\begin{cases} \Phi_e(0) = 0 \\ \frac{\partial \Phi_e(L_{\mathrm{cell}})}{\partial x} = 0 \end{cases}$
$C_{\mathrm{s,film}}^{\mathrm{DL}} a_s r_s \frac{\partial \Delta\Phi_{\mathrm{s,film}}(x)}{\partial t} = j_{\mathrm{s,film}}^{\mathrm{DL}}(x)$	-
$C_{\mathrm{ads,e}}^{\mathrm{DL}} a_s r_s \frac{\partial \Delta\Phi_{\mathrm{ads,e}}(x)}{\partial t} = j_{\mathrm{ads,e}}^{\mathrm{DL}}(x)$	-
$\frac{N_s}{o_s}\frac{\partial \theta^{m}(x)}{\partial t} = q_{\mathrm{ads}}^{m}(x)$	-

Table 4.2: Macroscopic and microscopic processes are considered for lithium de-/intercalation, degradation, sorption, and surface diffusion. Processes given as microscopic process are considered by the kMC model, while fluxes are synchronized with the continuum model, displayed as macroscopic processes to the left. Reproduced from publication [1] with kind permission from Wiley-VCH Verlag GmbH & Co. KGaA.

Nb.	Macroscopic Process	j	Microscopic process
I	$V(s) + e^-(s) + Li^+(ads) \rightleftharpoons Li(s) + V(ads)$	–	–
II	$Li^+(ads) + S_1(ads) + e^-(s) \rightleftharpoons LiS_1(ads) + V(ads)$	1	$Li^+(ads) + S_1(ads) + e^-(s) \rightarrow LiS_1(ads) + V(ads)$
		2	$LiS_1(ads) + V(ads) \rightarrow Li^+(ads) + S_1(ads) + e^-(s)$
III	$Li^+(ads) + S_2(ads) + e^-(s) \rightleftharpoons SEI_2(film) + 2V(ads)$	3	$Li^+(ads) + S_2(ads) + e^-(s) \rightarrow SEI_2(film) + 2V(ads)$
		4	$SEI_2(film) + 2V(ads) \rightarrow Li^+(ads) + S_2(ads) + e^-(s)$
IV	$Li^+(ads) + S_3(ads) + e^-(s) \rightleftharpoons SEI_3(film) + 2V(ads)$	5	$Li^+(ads) + S_3(ads) + e^-(s) \rightarrow SEI_3(film) + 2V(ads)$
		6	$SEI_3(film) + 2V(ads) \rightarrow Li^+(ads) + S_3(ads) + e^-(s)$
V	–	7	$2LiS_1(ads) \rightarrow SEI_1(film) + 2V(ads)$
		8	$SEI_1(film) + 2V(ads) \rightarrow 2LiS_1(ads)$
VI	$LiS_1(ads) \rightleftharpoons LiS_1(e) + V(ads)$	9	$LiS_1(ads) \rightarrow LiS_1(e) + V(ads)$
		10	$LiS_1(e) + V(ads) \rightarrow LiS_1(ads)$
VII	$2LiS_1(e) \rightleftharpoons SEI_1(e)$	–	–
VIII	$S_1(ads) \rightleftharpoons S_1(e) + V(ads)$	–	–
IX	$S_2(ads) \rightleftharpoons S_2(e) + V(ads)$	–	–
X	$S_3(ads) \rightleftharpoons S_3(e) + V(ads)$	–	–
XI	$Li^+(ads) \rightleftharpoons Li^+(e) + V(ads)$	–	–
XII	–	11	Horizontal surface diffusion: $LiS_1(ads) \rightarrow LiS_1(ads)$
XIII	–	12	Diagonal surface diffusion: $LiS_1(ads) \rightarrow LiS_1(ads)$

The applied current for a 1C charge is determined as

$$I_{\text{cell}}^{1\text{C}} = \frac{C_{\text{theo}}^{\text{As}}}{3600} \tag{4.1}$$

The theoretical capacity $C_{\text{theo}}^{\text{As}}$ is hereby defined as

$$C_{\text{theo}}^{\text{As}} = c_{\text{max}}^{a} \varepsilon_{s}^{a} \text{F} \delta_{a}. \tag{4.2}$$

using the anode thickness δ_a, the anode volume fraction ε_s^a, the maximal concentration in the anode c_{max}^a, and the Faraday constant F. Charge transfer at the interfaces is calculated as

$$j_{\text{s,film}}^{\text{ct}} = -a_s r_s \text{F}(q_{\text{I}} + q_{\text{II}} + q_{\text{III}} + q_{\text{IV}}) \tag{4.3}$$

for the interface between solid and film and

$$j_{\text{ads,e}}^{\text{ct}} = a_s r_s \text{F}(q_{\text{XI}}) \tag{4.4}$$

for the interface between electrolyte and adsorption layer. In these equations, a_s is the specific surface area, r_s is the roughness factor, and q are fluxes according to macroscopic processes given in Table 4.2. The specific surface area thereby is calculated as the surface area of spherical particles without roughness as

$$a_s = \frac{3\varepsilon_s}{R_s} \tag{4.5}$$

using the volume fraction of the solid material ε_s and the particle radius R_s. The current density of the electrochemical double layer can be calculated as

$$j_{\text{s,film}}^{\text{DL}} = j - j_{\text{s,film}}^{\text{ct}} \tag{4.6}$$

and

$$j_{\text{ads,e}}^{\text{DL}} = j - j_{\text{ads,e}}^{\text{ct}} \tag{4.7}$$

for ads/e and s/film interface, respectively. Here j is the current density. Calculation of the species' fluxes through an interface is shown at the example of flux of species S_1 through the ads/e interface:

$$q_{\text{ads,e}}^{\text{S}_{\text{I}}} = q_{\text{VII}}. \tag{4.8}$$

Accordingly, the balance for the net flux of S_1 at the adsorption site is applied as

$$q_{\mathrm{ads}}^{\mathrm{S_I}} = -q_{\mathrm{VII}} - q_{\mathrm{II}}. \tag{4.9}$$

Reaction fluxes are determined according to Colclasure et al. [117], which is shown in the following for the example of reaction I, i.e. an electrochemical reaction, and for the example of reaction VII, i.e. an adsorption process without charge being transferred between the interface, respectively:

$$\begin{aligned}\frac{o_s}{N_s} q_{\mathrm{I}} =& k_{\mathrm{I}}^{f} \theta^{\mathrm{V(ads)}} a^{\mathrm{Li}(s)} \exp\left(\frac{\beta \Delta\Phi_{\mathrm{a,film}} \mathrm{F}}{\mathrm{R}T}\right) \\ &- k_{\mathrm{I}}^{b} \theta^{\mathrm{Li^+(ads)}} a^{\mathrm{V(s)}} \exp\left(\frac{-(1-\beta) \Delta\Phi_{\mathrm{a,film}} \mathrm{F}}{\mathrm{R}T}\right)\end{aligned} \tag{4.10}$$

and

$$\frac{o_s}{N_s} q_{\mathrm{VII}} = k_{\mathrm{VII}}^{f} \theta^{\mathrm{S_1(ads)}} - k_{\mathrm{VII}}^{b} \theta^{\mathrm{V(ads)}} a_{\mathrm{S_1(e)}}, \tag{4.11}$$

Here, o_s is the surface site occupancy number, N_s is the surface site density, θ^m is surface fraction of species m, a^m is the activity of species m, β is the symmetry factor of the reaction, $\Delta\Phi$ is the electrical potential at the interface, R is the ideal gas constant, and T is the temperature. The surface fraction of vacancies on the surface is thereby calculated as

$$\begin{aligned}\theta_{\mathrm{V(ads)}} =& 1 - \theta_{\mathrm{Li^+(ads)}} - \theta_{\mathrm{S_1(ads)}} \\ &- \theta_{\mathrm{S_2(ads)}} - \theta_{\mathrm{S_3(ads)}} - \theta_{\mathrm{LiS_1(ads)}}\end{aligned} \tag{4.12}$$

The non ideal activity of lithium and vacancies in the solid material is determined as

$$a^{\mathrm{Li(s)}} = x^{\mathrm{Li}} \exp\left(\frac{(1-x^{\mathrm{Li}})^2}{RT} \sum_{i=0}^{N} \left(A_i^{\mathrm{RK}} (2x^{\mathrm{Li}} - 1)^i \left(1 + \frac{2ix^{\mathrm{Li}}}{2x^{\mathrm{Li}} - 1}\right)\right)\right) \tag{4.13}$$

and

$$a^{\mathrm{V(s)}} = (1 - x^{\mathrm{Li}}) \exp\left(\frac{(x^{\mathrm{Li}})^2}{RT} \sum_{i=0}^{N} \left(A_i^{\mathrm{RK}} (2x^{\mathrm{Li}} - 1)^i \left(1 - \frac{2i(1-x^{\mathrm{Li}})}{2x^{\mathrm{Li}} - 1}\right)\right)\right) \tag{4.14}$$

with A_i^{RK} being Redlich-Kister coefficients. The intercalation fraction x^{Li} is determined as

$$x^{\mathrm{Li}} = \frac{c^{\mathrm{Li(s)}}}{c_{\mathrm{max}}}. \tag{4.15}$$

Activity of species in the electrolyte is calculated assuming ideal solution as

$$a^m = \frac{c^m}{C_m^0}, \tag{4.16}$$

with C^0 being the standard state concentration.

Forward and backward reaction rate constant are calculated, as shown at the example of reaction I, as Arrhenius expressions

$$k_{\mathrm{I}}^f = A \exp\left(\frac{-E_{\mathrm{I}}^A}{\mathrm{R}T}\right) \tag{4.17}$$

and

$$k_{\mathrm{I}}^b = A \exp\left(\frac{-(E_{\mathrm{I}}^A - \Delta G_{\mathrm{I}}^0)}{\mathrm{R}T}\right) \tag{4.18}$$

with A being the pre exponential factor, E^A being the activation energy of the forward reaction, and $E^A - \Delta G^0$ being the activation energy of the backward reaction. For the sake of simplicity and as no dependency of T is considered, A is chosen equally for all processes. Thereby ΔG^0 is the standard state Gibbs free energy of the reaction, which can be determined as

$$\Delta G_{\mathrm{I}}^0 = \mu_{\mathrm{Li(s)}}^0 + \mu_{\mathrm{V(ads)}}^0 - \mu_{\mathrm{V(s)}}^0 - \mu_{\mathrm{Li^+(ads)}}^0 \tag{4.19}$$

using the standard state chemical potential μ^0 of the involved species. Applied standard state chemical potentials are summarized in the appendix A in Table A.6.

To simulate transport processes in the electrolyte within the porous electrode, effective diffusion coefficients are determined as

$$D_{e,\mathrm{eff}}^m(x) = \frac{\varepsilon_e(x)}{\tau_e} D_e^m \tag{4.20}$$

with D_e^m being the bulk diffusion coefficient of species m, ε_e being the electrolyte volume fraction, and τ_e being the tortuosity within the electrolyte. Surface films pose an additional resistance for lithium-ion transport, but they also reduce the porosity of the electrode and hinder transport processes in the electrolyte, as shown by [32]. Therefore, the electrolyte volume fraction is calculated as

$$\varepsilon_e(x) = 1 - \varepsilon_s - a_s r_s d^{\mathrm{film}}(x) \tag{4.21}$$

with d^{film} being the average thickness of the surface film. Thus, increasing the film thickness will reduce the porosity of the electrode and affect transport processes on the electrode scale. Diffusion coefficients are assumed to be constant for solvent and intermediate species and determined as concentration dependent for lithium, as significant concentration gradients for lithium salt during operation can be expected. The concentration dependency is described as a polynomial

$$D_e^{\text{salt}} = b_1 + b_2 c_e^{\text{salt}} + b_3 {c_e^{\text{salt}}}^2 \tag{4.22}$$

using the coefficients b_1–b_3. Since no detailed experimental data for concentration dependency of the diffusion coefficient is available for this work, the diffusion coefficient is directly linked to ionic conductivity using Nernst-Einstein relation. We note that this is usually restricted to dilute solutions. Nevertheless, it is applied here to provide a qualitative feasible concentration dependency for ionic conductivity and diffusion coefficient using only experimental data for ionic conductivity. The ionic conductivity σ_e is defined as a function of the salt diffusion coefficient:

$$\sigma_e(x) = c_e^{\text{salt}}(x) \frac{\text{F}^2 D_e^{\text{salt}}(c_e^{\text{salt}})}{t_p(2 - 2t_p)\text{R}T}. \tag{4.23}$$

According to Legrand et al. [134], the diffusional ionic transport coefficient is determined as

$$\sigma_{\text{De}} = \frac{2\text{R}T(t_p - \frac{1}{2}))}{\text{F}} \sigma_e \tag{4.24}$$

Effective transport coefficients in the solid and electrolyte phase are determined as

$$\sigma_{s,\text{eff}} = \varepsilon_s \sigma_s \tag{4.25}$$

and

$$\sigma_{e,\text{eff}}(x) = \frac{\varepsilon_e(x)}{\tau_e} \sigma_e(x), \tag{4.26}$$

respectively. The difference between electrical potential in the solid and electrolyte phase is determined as

$$\begin{aligned} \Delta\Phi_{\text{s,e}}(x) =& \Phi_s(x) - \Phi_e(x) \\ =& \Delta\Phi_{\text{s,film}}(x) + \Delta\Phi_{\text{ads,e}}(x) + \frac{j(x) R^{\text{film}} d^{\text{film}}(x)}{a_s r_s} \end{aligned} \tag{4.27}$$

and includes the potential drop at the s/film interface, the ads/e interface, and ohmic losses within the surface film, which depends on the current density j, average film thickness d^{film}, and the specific resistance of the film R^{film}.

4.3.2 Identification of parameters of operational processes

Parameters applied within this chapter can be found in the appendix A in Table A.8. Most geometrical parameters, i.e. electrode and separator thicknesses, solid volume fraction, and particle sizes, were measured or taken from material data sheets. The main thermodynamic and kinetic parameters were identified as summarized in the following.

First, the thermodynamic parameters of the main intercalation reaction were determined. The Redlich-Kister coefficients and the standard state chemical potentials of lithium in the solid $\mu^0_{\mathrm{Li(s)}}$ can be identified by measuring equilibrium potential, i.e. OCP, of the electrodes against a lithium reference, as given in Figure 4.2, and adapting the coefficients to minimize the difference from the calculated equilibrium potential vs Li metal, which is

$$E^{\mathrm{eq}}_{\mathrm{ref}} = \frac{\mu^0_{\mathrm{V(s)}} + \mu^0_{\mathrm{Li,metal}} - \mu^0_{\mathrm{Li(s)}}}{\mathrm{F}} + \frac{\mathrm{R}T}{\mathrm{F}} \ln\left(\frac{a^{\mathrm{V(s)}}}{a^{\mathrm{Li(s)}}}\right). \tag{4.28}$$

The calculated potential is compared to measured values in Figure 4.2. The identified Redlich-Kister coefficients are provided in the appendix A in Table A.7 and the identified standard state chemical potentials of lithium in the solid $\mu^0_{\mathrm{Li(s)}}$ is provided in the appendix A in Table A.6.

Secondly, transport properties of the electrolyte with respect to the main processes can be evaluated by measuring electrical conductivity. To achieve a quantitative agreement of electrical conductivity, coefficients b_1–b_3 have been adapted to achieve a good agreement to a data point taken from Plichta et al. [136] and to reproduce the typical nonmonetonous concentration dependency of lithium-ion battery electrolytes, as provided in [137]. The applied concentration dependency is shown in Figure 4.3. This yields a good quantitative description of the electrolyte conductivity, which is the main feature of the electrolyte. However, future work should refine parameter identification of the electrolyte and include measurement of the concentration dependency of diffusion coefficient and transference number in concentrated solutions.

Finally, the main kinetic parameters, i.e. activation energies, $E_1^{A,a}$ and $E_1^{A,c}$, solid diffusion coefficients, $D_s^{\mathrm{Li},a}$ and $D_s^{\mathrm{Li},c}$, and electrical conductivity within the solid, σ_s^a and σ_s^c, as well as unknown geometrical properties such as tortuosity, τ_e^a and τ_e^c, are identified using C-rate tests as shown in Figure 4.4. It should be noted that the parameter identification, as shown here, does not provide proof for the uniqueness of the identified parameter set. Moreover, as activation energies have been parameterized only for one temperature, they only provide a feasible reaction rate for this temperature but do not accurately describe actual temperature dependency. However, good agreement between simulated and measured electrical potentials was achieved.

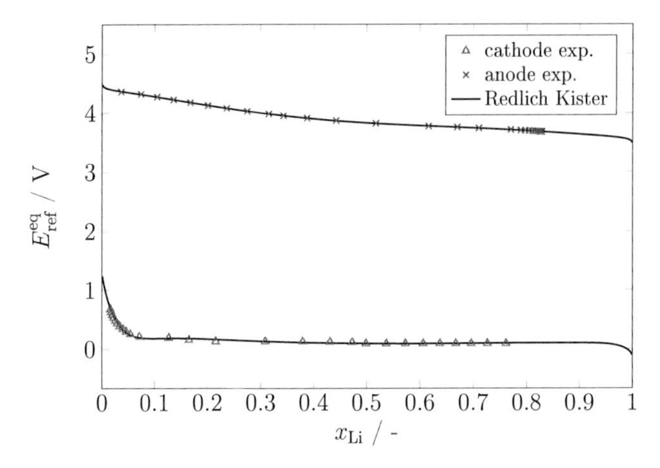

Figure 4.2: Measured and calculated equilibrium potential of anode and cathode against a lithium metal reference. Reproduced from publication [1] with kind permission from Wiley-VCH Verlag GmbH & Co. KGaA.

In the case of the multiscale simulation, the continuum model relies on kMC input parameters, i.e. $\theta_{\mathrm{LiS_1(ads)}}$, q_{II}, q_{III}, q_{IV}, q_{VI}, and d^{film}, which is shown in Figure 4.1. The reaction rate fluxes q need to be considered if the contribution of side reactions is significant. This is the case during the first charge/discharge cycles of the battery. However, single discharge cycles after formation are assumed to have negligible progress in degradation, i.e. the degradation state is set to quasi static state. The assumption has been applied to identify the parameters of the operational processes, i.e. normal charge/discharge behavior of the battery. In detail, for parameter identification shown in this section, it is assumed that $\theta_{\mathrm{LiS_1(ads)}} = 0$, $q_{\mathrm{II}} = 0$, $q_{\mathrm{III}} = 0$, $q_{\mathrm{IV}} = 0$, $q_{\mathrm{VI}} = 0$, and $d^{\mathrm{film}} = 4 \times 10^{-8}$ m, i.e. corresponding to no side reactions and an assumed film thickness of 40 nm. For the purpose of this work, the initial guess was sufficiently precise. However, in future work film thickness can be adapted after simulation of formation in an iterative process, which would refine the parameter identification procedure. Assumed film thickness could be adapted after simulation of formation in an iterative process. However, for the purpose of this chapter, the initial guess was sufficiently precise. Identifying parameters of the operational processes a priori to the analysis of the first charge cycle, allows for the separation of the contribution of the operational processes during the first charge cycle and thus enables the parameter identification of the side reactions.

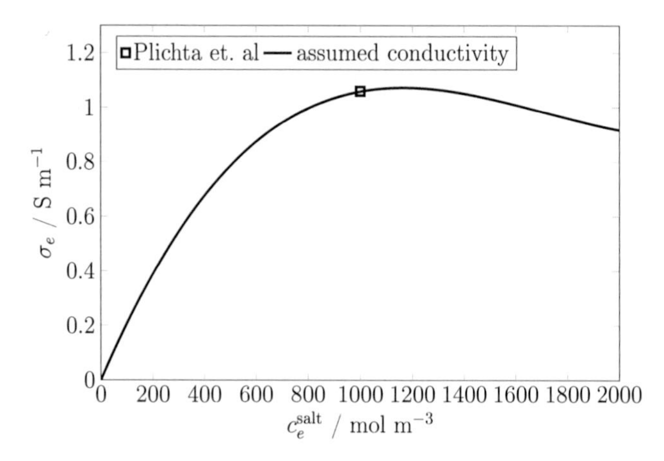

Figure 4.3: Electrolyte conductivity as a function of salt concentration and measurment point taken from Pilcha et al. [136]. Reproduced from publication [1] with kind permission from Wiley-VCH Verlag GmbH & Co. KGaA.

4.4 Microscopic scale

4.4.1 Kinetic Monte Carlo model

In order to model the detailed microscopic processes at the film surface and the heterogeneous film growth, the continuum model, as shown in the previous section, is directly coupled with a kMC model. Details can be found in in chapter 2 and 3.

In every time step i of the kMC model, the kMC algorithm performs the following actions:

1. calculate microscopic rate, $\Gamma_i^{l,j}$ for every possible microscopic process $j \in \{z|z \in \mathbb{N}, z \leq n^j\}$ on every lattice site $l \in \{z|z \in \mathbb{N}, z \leq n^l\}$.

2. calculate time step, Δt.

3. select one process $J_i \in \{z|z \in \mathbb{N}, z \leq n^j\}$ and lattice site $L_i \in \{z|z \in \mathbb{N}, z \leq n^l\}$ according to microscopic rates $\Gamma_i^{l,j}$ (details below).

4. perform process J_i.

5. if end time of simulation not reached go to 1

The model is set up as a solid-on-solid, i.e. 2+1D, model. The model structure is illustrated in Figure 4.1 (C). The state of the film sites f in a 3D cubic system is

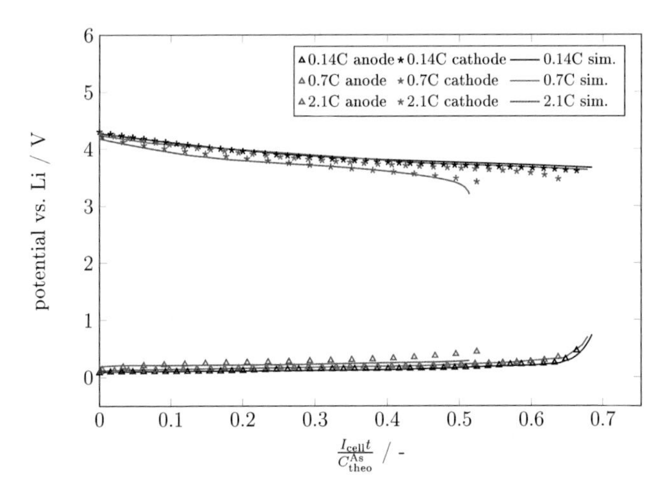

Figure 4.4: Measurement and simulation of a C-rate test for anode and cathode potential versus lithium reference placed in the center of the separator. Reproduced from publication [1] with kind permission from Wiley-VCH Verlag GmbH & Co. KGaA.

defined as

$$\vartheta_i^{\mathrm{film}}(f) \in \{\mathrm{solid}, \mathrm{SEI}_1, \mathrm{SEI}_2, \mathrm{SEI}_3, \vartheta_i^{\mathrm{lattice}}\} \tag{4.29}$$

The bottom row of the 3D film is defined to be filled with solid, e.g. anode or cathode active material; the other sites can be either a SEI component, or a lattice site, where species are adsorbed. Film growth is restricted to vertical growth with respect to the solid surface. The lattice site is defined as being on top of the highest solid component, as illustrated in Figure 4.1 (C). Microscopic processes on the surface are considered to be on a 2D lattice, with lattice site l. Neighbors of a given film site f are indicated as f'. Horizontal neighbors of a lattice site l are indicated as l' and diagonal neighbors are indicated as l''. A film site corresponding to a lattice site l is indicated as $f(l*)$, and film site directly below a lattice site are indicated as $f(l)$. There are two lattice site states

$$\vartheta_i^{\mathrm{lattice}}(l) \in \{\neg\mathrm{LiS}_1, \mathrm{LiS}_1\} \tag{4.30}$$

where LiS_1 corresponds to a surface site covered with a LiS_1 species, and $\neg\mathrm{LiS}_1$ corresponds to a surface not covered with a LiS_1 species, i.e. possibly covered with other species or being vacant. The probability that a $\neg\mathrm{LiS}_1$ site is covered with a species m is determined as

$$\theta^{m}_{\neg\text{LiS}_1} = \frac{\theta^{m}}{1 - \theta^{\text{LiS}_1}} \tag{4.31}$$

The direct inclusion of only a few species in the kMC model, with other species being covered by the continuum model, enables a long time length for multiscale simulation and has been shown in detail in chapter 2.

In every kMC step i, the microscopic rate $\Gamma_i^{l,j}$ of every considered microscopic process j and every lattice site l is calculated. Considered microscopic processes j are given with Table 4.2. The table also shows the corresponding macroscopic processes. In Table 4.3 assumed requirements, i.e. for solids and species on neighboring sites, the performed action in case of selection, and equations for calculation of the microscopic rates are given.

Table 4.3: Microscopic processes j within the kMC model, according to Table 4.2 and the corresponding lattice and film site states before and after the process. Reproduced from publication [1] with kind permission from Wiley-VCH Verlag GmbH & Co. KGaA.

j	state at i	state at $i+1$	microscopic rate $\Gamma_i^{l,j}$
1	$\begin{cases} \vartheta_i^{\text{lattice}}(l) = \neg\text{LiS}_1 \\ \vartheta_i^{\text{lattice}}(l') = \neg\text{LiS}_1 \end{cases}$	$\vartheta_{i+1}^{\text{lattice}}(l) = \text{LiS}_1$	$k_{\text{II}}^{f}[\theta^{\text{Li}^+}_{\neg\text{LiS}_1}][\theta^{\text{S}_1}_{\neg\text{LiS}_1}]p_{\text{cat}}$
2	$\begin{cases} \vartheta_i^{\text{lattice}}(l) = \text{LiS}_1 \\ \vartheta_i^{\text{lattice}}(l') = \neg\text{LiS}_1 \end{cases}$	$\vartheta_{i+1}^{\text{lattice}}(l) = \neg\text{LiS}_1$	$k_{\text{II}}^{b}[\theta^{\text{V}}_{\neg\text{LiS}_1}]p_{\text{an}}p_{\text{bond},f(l^*)}$
3	$\begin{cases} \vartheta_i^{\text{lattice}}(l) = \neg\text{LiS}_1 \\ \vartheta_i^{\text{lattice}}(l') = \neg\text{LiS}_1 \end{cases}$	$\vartheta_{i+1}^{\text{film}}(f(l^*)) = \text{SEI}_2$	$k_{\text{III}}^{f}[\theta^{\text{Li}^+}_{\neg\text{LiS}_1}][\theta^{\text{S}_2}_{\neg\text{LiS}_1}]p_{\text{cat}}$
4	$\begin{cases} \vartheta_i^{\text{lattice}}(l) = \neg\text{LiS}_1 \\ \vartheta_i^{\text{lattice}}(l') = \neg\text{LiS}_1 \end{cases}$	$\vartheta_{i+1}^{\text{film}}(f(l)) = \vartheta_{i+1}^{\text{lattice}}(l)$	$k_{\text{III}}^{b}[\theta^{\text{V}}_{\neg\text{LiS}_1}]^2 p_{\text{an}}p_{\text{bond},f(l)}$
5	$\begin{cases} \vartheta_i^{\text{lattice}}(l) = \neg\text{LiS}_1 \\ \vartheta_i^{\text{lattice}}(l') = \neg\text{LiS}_1 \end{cases}$	$\vartheta_{i+1}^{\text{film}}(f(l^*)) = \text{SEI}_3$	$k_{\text{IV}}^{f}[\theta^{\text{Li}^+}_{\neg\text{LiS}_1}][\theta^{\text{S}_3}_{\neg\text{LiS}_1}]p_{\text{cat}}$
6	$\begin{cases} \vartheta_i^{\text{lattice}}(l) = \neg\text{LiS}_1 \\ \vartheta_i^{\text{lattice}}(l') = \neg\text{LiS}_1 \end{cases}$	$\vartheta_{i+1}^{\text{film}}(f(l)) = \vartheta_{i+1}^{\text{lattice}}(l)$	$k_{\text{IV}}^{b}[\theta^{\text{V}}_{\neg\text{LiS}_1}]^2 p_{\text{an}}p_{\text{bond},f(l)}$
7	$\begin{cases} \vartheta_i^{\text{lattice}}(l) = \text{LiS}_1 \\ \vartheta_i^{\text{lattice}}(l') = \text{LiS}_1 \end{cases}$	$\begin{cases} \vartheta_{i+1}^{\text{lattice}}(l) = \neg\text{LiS}_1 \\ \vartheta_{i+1}^{\text{lattice}}(l') = \neg\text{LiS}_1 \\ \vartheta_{i+1}^{\text{film}}(f(l^*)) = \text{SEI}_1 \\ \vartheta_{i+1}^{\text{film}}(f(l'^*)) = \text{SEI}_1 \end{cases}$	k_{V}^{f}
8	$\begin{cases} \vartheta_i^{\text{lattice}}(l) = \neg\text{LiS}_1 \\ \vartheta_i^{\text{lattice}}(l') = \neg\text{LiS}_1 \\ \vartheta_i^{\text{film}}(f(l)) = \text{SEI}_1 \\ \vartheta_i^{\text{film}}(f(l')) = \text{SEI}_1 \end{cases}$	$\begin{cases} \vartheta_{i+1}^{\text{lattice}}(l) = \text{LiS}_1 \\ \vartheta_{i+1}^{\text{lattice}}(l') = \text{LiS}_1 \\ \vartheta_{i+1}^{\text{film}}(f(l)) = \vartheta_{i+1}^{\text{lattice}}(l) \\ \vartheta_{i+1}^{\text{film}}(f(l')) = \vartheta_{i+1}^{\text{lattice}}(l') \end{cases}$	$k_{\text{V}}^{b}[\theta^{\text{V}}_{\neg\text{LiS}_1}]^2 p_{\text{bond},f(l)}p_{\text{bond},f(l')}$
9	$\vartheta_i^{\text{lattice}}(l) = \text{LiS}_1$	$\vartheta_{i+1}^{\text{lattice}}(l) = \neg\text{LiS}_1$	$k_{\text{VI}}^{f}p_{\text{bond},f(l^*)}$
10	$\vartheta_i^{\text{lattice}}(l) = \neg\text{LiS}_1$	$\vartheta_{i+1}^{\text{lattice}}(l) = \text{LiS}_1$	$k_{\text{VI}}^{b}[\theta^{\text{V}}_{\neg\text{LiS}_1}][a_{\text{LiS}_1(e)}]$
11	$\begin{cases} \vartheta_i^{\text{lattice}}(l) = \text{LiS}_1 \\ \vartheta_i^{\text{lattice}}(l') = \neg\text{LiS}_1 \end{cases}$	$\begin{cases} \vartheta_{i+1}^{\text{lattice}}(l) = \neg\text{LiS}_1 \\ \vartheta_{i+1}^{\text{lattice}}(l') = \text{LiS}_1 \end{cases}$	$\frac{D_{\text{asd}}^{\text{LiS}_1}}{2\Delta L^2}p_{\text{bond},f(l^*)}$
12	$\begin{cases} \vartheta_i^{\text{lattice}}(l) = \text{LiS}_1 \\ \vartheta_i^{\text{lattice}}(l'') = \neg\text{LiS}_1 \end{cases}$	$\begin{cases} \vartheta_{i+1}^{\text{lattice}}(l) = \neg\text{LiS}_1 \\ \vartheta_{i+1}^{\text{lattice}}(l'') = \text{LiS}_1 \end{cases}$	$\frac{D_{\text{asd}}^{\text{LiS}_1}}{4\Delta L^2}p_{\text{bond},f(l^*)}$

Within the kMC algorithm the time step length is calculated based on a uniformly distributed random number $\zeta_1 \in (0,1)$ as

$$\Delta t_{i+1}^{\mathrm{kMC}} = \frac{-\ln(\zeta_1)}{\Gamma_i^{\mathrm{tot}}}, \tag{4.32}$$

thus the time at the following time step $i+1$ can be calculated with

$$t_{i+1}^{\mathrm{kMC}} = t_i^{\mathrm{kMC}} + \Delta t_{i+1}^{\mathrm{kMC}}. \tag{4.33}$$

Further, in every kMC step, one of the possible microscopic processes is selected using a second uniformly distributed random number $\zeta_2 \in (0,1)$ according to

$$\frac{\sum_{k=1}^{L_i}\sum_{j=1}^{J_i-1}\Gamma_i^{k,j}}{\Gamma_i^{\mathrm{tot}}} < \zeta_2 \leq \frac{\sum_{k=1}^{L_i}\sum_{j=1}^{J_i}\Gamma_i^{k,j}}{\Gamma_i^{\mathrm{tot}}} \tag{4.34}$$

with J_i being the selected process and L_i being the selected lattice site. The total microscopic rate Γ_i^{tot} is calculated as

$$\Gamma_i^{\mathrm{tot}} = \sum_{k=1}^{n^k}\sum_{j=1}^{n^J}\Gamma_i^{k,j}. \tag{4.35}$$

A boolean, which indicates the selection of a process j within a kMC step i can be determined as

$$\Psi_i^j = \begin{cases} 1, & \text{if } J_i = j \\ 0, & \text{otherwise} \end{cases} \tag{4.36}$$

The single processes and equations for calculating the corresponding microscopic rates $\Gamma_i^{l,j}$ are given in Table 4.3. These rates can depend on thickness of the film and breakage of bonds to nearest neighbors, as explained in the following.

Several growth limiting steps are assumed in literature, e.g. electron tunneling, electron conduction, solvent diffusion or lithium-interstitial diffusion, which have been studied using continuum models for aging during battery storage [138] or formation [139]. Possibly there is a competition of several transport mechanisms in parallel [140] and a two layer structure with a dense inner and porous outer layer [141]. Shi et al. [84] identified detailed lithium and electron transport properties within Li_2CO_3 using DFT simulations, which also indicates that the transport mechanism depend on various factors, such as molecular structure, potential, and active material. Tang et al. [140] provided a detailed study to identify growth limiting steps during formation, but also conclude that the transport mechanism can depend on the electrode structure and may differ between formation and capacity fade.

In this work, the probability of an electron leaking through the surface film is assumed to exponentially decrease with film thickness. This is usually associated to electron tunneling, and has been applied frequently, e.g. [28, 142]. This assumption provides good agreement to experiments shown in this work. It is noted that as outlined above, the actual mechanism strongly depends on the SEI composition and structure. Further, limitation of film growth originating from film porosity is not considered, but only the impact of reduced porosity on electrode scale. Identification of mechanisms of growth limitation are not within the scope of this work and thus, actual mechanisms may differ and can be significantly more complex.

The following equations for cathodic and anodic reactions, respectively:

$$p_{\mathrm{cat},l} = \exp\left(\frac{-(1-\beta)\Delta\Phi\mathrm{F} - \hat{d}^{\mathrm{film}}(l)\hat{E}^{A}_{\mathrm{film}}}{\mathrm{R}T}\right) \tag{4.37}$$

and

$$p_{\mathrm{an},l} = \exp\left(\frac{\beta\Delta\Phi\mathrm{F} - \hat{d}^{\mathrm{film}}(l)\hat{E}^{A}_{\mathrm{film}}}{\mathrm{R}T}\right). \tag{4.38}$$

with $\hat{d}^{\mathrm{film}}(l)$ being the film thickness at the lattice site l. Further, the probability of a bond breakage between two neighboring sites f and f' is calculated according to [18] as

$$p_{\mathrm{bond},f} = \exp\left(\frac{-\sum_{f'} E^{A}_{\mathrm{bond}}(\vartheta^{\mathrm{film}}(f)|\vartheta^{\mathrm{film}}(f'))}{\mathrm{R}T}\right) \tag{4.39}$$

The binding of a species is illustrated with a simple example in Figure 4.5. Here a first a species A is adsorbed at the surface as B. Thereby, species B binds to the surface site, which reduces its chemical energy. This denotes that for a desorption step or a reaction step, i.e. B→C, the activation energy of this processes and the activation energy for bond breakage is considered.

The output of the kMC model, i.e. input parameters to the continuum model, is provided in the following equations. The average thickness of the surface film is determined as

$$d^{\mathrm{film}} = \frac{\sum_l \hat{d}^{\mathrm{film}}}{n^l} \tag{4.40}$$

and the reaction fluxes q are determined as shown at the example of reaction II as

$$q_{\mathrm{II}} = \frac{\sum_{\mathrm{seq}} \Psi_i^{j=1} - \sum_{\mathrm{seq}} \Psi_i^{j=2}}{\Delta t_{\mathrm{seq}} n^l \Delta L^2 N_s}. \tag{4.41}$$

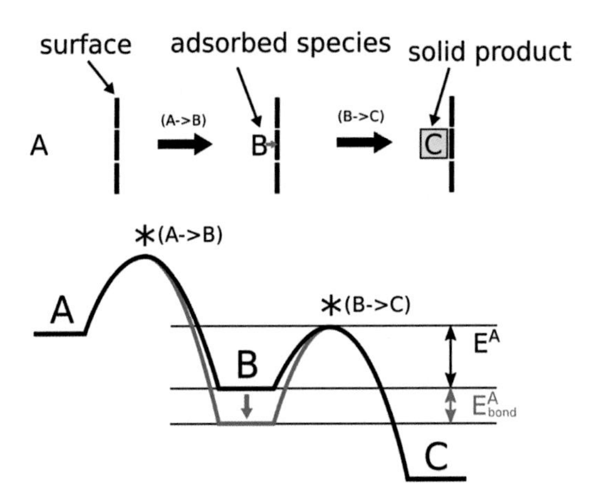

Figure 4.5: Illustration of surface binding at a simple example: A→B, with B binding to the surface, and B→C.

Further, the average surface fraction of species LiS_1 is determined. Output data is evaluated within a sequence (seq). Details about sequences and model coupling can be found in chapter 2. Within the multiscale simulations, information is sequentially exchanged between the kMC and continuum models. Exchanged data is summarized in Figure 4.1.

4.4.2 Identification of parameters of degradation process

The applied experiments are not particularly designed to identify parameters of complex surface film growth mechanisms, but are rather standard characterization experiments. Further, presently neither adequate experimental nor theoretical data is available to describe decomposition reactions for an electrolyte with this complexity. Therefore, instead, a lumped degradation mechanism is constructed, which (i) includes comprehensive heterogeneity, but at the same time is (ii) simple enough to generate illustrative results with traceable multiscale effects, and (iii) enables one to identify parameters that achieve quantitative agreement with the experimental data. The presented results do not claim to provide details about degradation chemistry, but instead demonstrate the simulation method and provide evidence for multiscale effects. In the following, the constructed mechanism, motivation for the assumptions are outlined.

As given in Section 4.2.1, experiments were carried out using a tertiary electrolyte with three solvents, i.e. EC, DMC, and EMC. Components are electrochemically unstable in the applied potential window and thus at least three different SEI components can be expected. Thus, for the presented analysis, a tertiary electrolyte with solvents S_1, S_2 and S_3 is assumed.

Rate constants and decomposition potentials vary for different solvents [143]. Potentials for decomposition are reported to be between 0.6-1 V [144, 109]. Xu et al [145] indicate that solvents have different lowest unoccupied molecule orbitals, which has a high impact on reaction ability and thus reduction potential. For instance they showed that the value of EC is much lower than the one of EMC, which denotes that theoretically EC decomposes at higher potentials than EMC. In the presented experimental data for the first charge cycle, as given in Figure 4.6, the decomposition reaction starts at 0.8 V. Thus decompostion potential of SEI_1 and SEI_2 is chosen with 0.8 V. To investigate the effect of the outlined variation in decomposition potential for different solvents a potential of 0.6 V is chosen in case of SEI_3.

Agabura et al. [126] suggest various mechanism for solvent decomposition. For instance they suggest that EMC may decompose in a two electron process and two lithium ions to form lithium carbonate. An et al. [109] suggest that EC possibly first reacts with e^- and Li^+, which in a second step further decomposes to form for instance lithium ethylene dicarbonate. As mechanisms are still under discussion and often several possible pathways are suggested, here the following lumped mechanisms are assumed:

$$S_1 + e^- + Li^+ \rightleftharpoons LiS_1 \tag{4.42}$$

$$2\ LiS_1 \rightleftharpoons SEI_1 \tag{4.43}$$

,

$$S_2 + e^- + Li^+ \rightleftharpoons SEI_2 \tag{4.44}$$

$$S_3 + e^- + Li^+ \rightleftharpoons SEI_3 \tag{4.45}$$

As a standard case the most simple possible decomposition mechanism is chosen, where a solvent reacts with Li^+ and e^- to form a SEI component. This mechanism is chosen for decomposition of S_2 and S_3. For S_1 a slightly more complex mechanism is chosen, where first an intermediate, LiS_1, is formed by reaction of S_1 with a Li^+ and an e^-. LiS_1 diffuses on the surface and reacts with neighboring species on the surface to form the SEI_1. With this, the decomposition of S_1 includes two reaction steps. The second step, i.e. Equation 4.43, does not involve an electron transfer and thus can also take place within the electrolyte solution after LiS_1 desorption.

Since the mechanism does not include any electrode specific components, consistency requires consideration for both electrodes, i.e. anode and cathode. However, due to the high potential no film formation takes place at the cathode.

As outlined previously, SEI is often found to have heterogenous structure, which can be caused by crystalline growth [108] or formation of clusters [131]. In this chapter heterogeneity is introduced as explained in the following. It is assumed that the intermediate LiS_1 has stronger binding energies to its decomposition product SEI_1. Further, it is assumed that the chemical energy of the SEI_1 decreases with increasing number of bonds. This denotes that large clusters are favorable. Finally, we assume that SEI_2 and SEI_3 have repulsive bonds, which denotes the chemical energy is increased, i.e. activation energy for reaction is decreased, if they are on neighboring sites.

In Figure 4.6, the measured potential during the first charge of the battery is presented. Further, the simulation results for this charge process using the continuum model, i.e. only Li intercalation, is shown as dashed red line. It can be seen that there is a significant deviation between measurement and simulation. Since the model is in good agreement with the experiments after film formation, this deviation can be assigned to the side reactions, i.e. film formation. It can be seen that at a certain potential, i.e. approximately 0.8 V, there is a small potential plateau. Further, the potential at the beginning of charge is significantly higher. The deviation corresponds to the electrical charge, which is consumed by the side reactions in addition to the operational process of lithium intercalation.

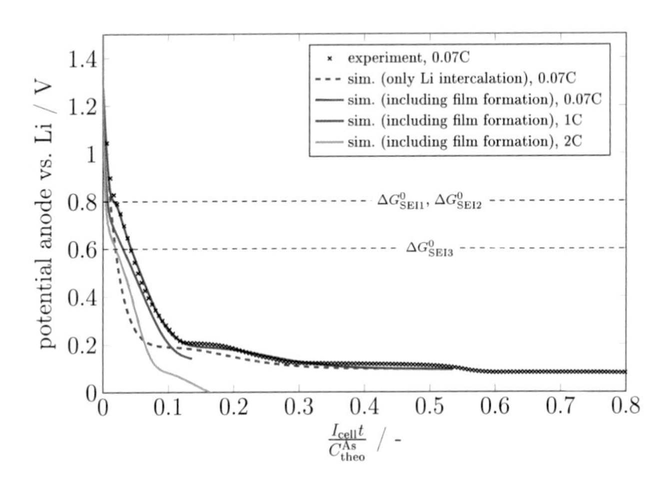

Figure 4.6: Potential during first charge, i.e. film formation, of battery with 0.07C for experimental and 0.07C, 1C, and 2C for the simulation. Reproduced from publication [1] with kind permission from Wiley-VCH Verlag GmbH & Co. KGaA.

The height of the observed plateau corresponds to the Gibbs free energy between educts and products of the single SEI reaction and thus triggers the respective reactions at potentials below the corresponding potential. The potential has been set by choosing respective values for the standard state chemical potential μ^0 of the SEI components in the surface film. The length of the plateau mainly corresponds to the time which is needed to form a stable initial film, i.e. the first depletion or desorption of solid SEI components, after which rates of side reactions are reduced according to Equations 4.37 and 4.38. The slope at the beginning of the first discharge mainly corresponds to the specific activation energy for electron leakage $\hat{E}^A$. Based on these findings, parameters $\mu^0_{\mathrm{SEI_1(ads)}}$, $\mu^0_{\mathrm{SEI_2(ads)}}$, and $\hat{E}^A$ have been selected to achieve agreement between the multiscale simulation and the experiment for the first charge of the battery with 0.07C, which is shown in Figure 4.6.

In Figure 4.6, further, potential is shown for higher C-rates, i.e. 0.5C, 1C, and 2C. Goers et al. [144] presented experimental results for formation with variation of the applied current. Their experiments are qualitatively in agreement with the presented simulation results for electrode potential during the first charge. They observed, that the formation plateau "remained almost constant with varying overall current density" and it "shifted with increasing current density towards more negative potentials".

Note that the chosen decomposition mechanisms, i.e. Equation 4.42–4.45, are lumped and most of the parameters of the underlying microscopic processes remain unknown. Thus, the results will not provide details about the actual mechanism within the analyzed cell. Future work will be needed to refine the parameterisation process and allow more detailed kinetic studies. However, the choice of these mechanisms and parameters fulfill the conditions i-iii as stated at the beginning of this section. Key parameters corresponding to the formation process have been selected to achieve good agreement between the measured and simulated potential during film formation, which denotes that the quantity of the side reaction is validated. Therefore simulation results shown here are valid for an analysis of general aspects of interaction between main intercalation, side reactions, and heterogeneous film growth, which is the scope of this paper and discussed in the following section.

4.5 Results and discussion

In the previous sections, the model and a general procedure to adjust parameters to fit experimental data has been outlined. In this section, a parmeterized model is used to provide evidence for multiscale effects, which with a high likelihood occur in a similar manner in real lithium-ion batteries. Simulations are quantitatively in very good agreement with the presented experiments. This denotes that the magnitude of SEI components produced, as well as the potentials of the decomposition reactions, are plausible. On this basis, no conclusions about the actual cell chemistry and film structure can be made, but instead the general aspects of the multiscale interaction and their potential impact on film structure can be revealed to illustrate the current dependence of multiscale effects.

Additionally to the experimentally applied formation procedure, i.e. 0.07C, 2.15 A m^{-2}, simulations are performed with 1C, 30.7 A m^{-2}, and 2C, 61.4 A m^{-2}, which are considerably faster procedures. In the following, first, the major observations of the simulations are outlined, and afterwards the multiscale interaction is discussed.

4.5.1 Simulation results

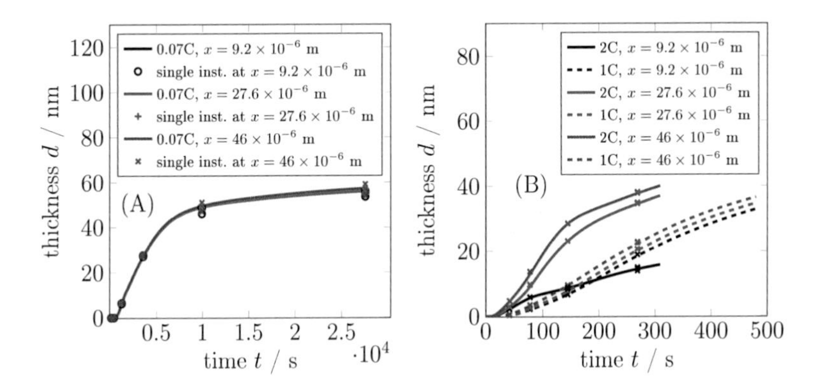

Figure 4.7: Film thickness at three positions x in the anode for simulation of formation with (A) 0.07C and (B) 1C and 2C. Further the result for single kMC instances at characteristic time points are given to elucidate statistical fluctuations. Reproduced from publication [1] with kind permission from Wiley-VCH Verlag GmbH & Co. KGaA.

In Figure 4.7 (A) average thickness, out of 9 simulated kMC instances, of surface film for formation with 0.07C at three different positions x within the electrode is shown, while small values of x corresponding to positions close to the current collector. It can be seen, that film thickness monotonously increases with time. Film growth rate decreases at about 1×10^4 s. This strongly correlates to the potential change at this time shown in Figure 4.6. It is noted that this correlation is mainly related to the defined electron leakage mechanism. This is usually associated with electron tunneling, which should be limited to a few nanometers. With significantly greater thicknesses than 5 nm the leakage mechanism may change, which could yield a deviating slope. Further, it can be seen that there is a small difference in film thickness along the x coordinate with higher film thickness close to the separator. Further, the values for film thickness of four single instances of the kMC model are given. It can be seen that there is a significant fluctuation in thickness of up to approximately 3 nm, which is increasing with increasing time t and which is considerably larger than the differences along electrode position x.

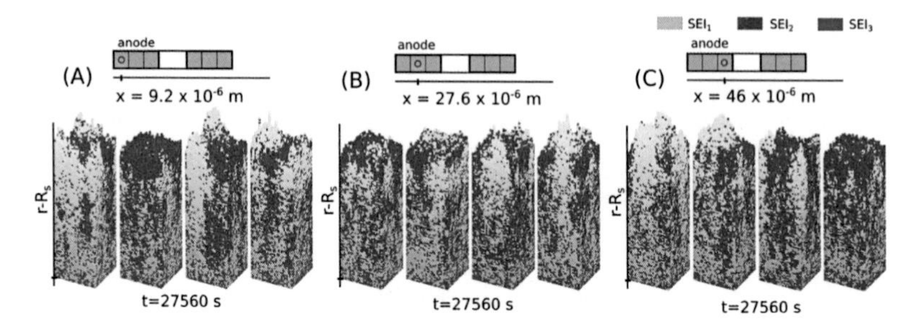

Figure 4.8: Film structure after film formation for four kMC instances with equivalent input parameters at 0.07C, $t = 27560$ s, and position (A) $x = 9.2 \times 10^{-6}$ m, close to current collector, (B) $x = 26.6 \times 10^{-6}$ m, center of electrode, (C) $x = 46 \times 10^{-6}$, close to separator. Reproduced from publication [1] with kind permission from Wiley-VCH Verlag GmbH & Co. KGaA.

In Figure 4.8 (A–C), the corresponding molecular structure of the surface film after film formation is shown at the same three positions within the electrode for the four parallel kMC instances displayed in Figure 4.7 (A) at the end of formation. It can be seen that the assumed decomposition mechanisms indeed yield significant heterogeneity in film structures. Moreover, it can be seen that the four parallel instances deviate due to the stochastic nature of the kMC model. There is a significant stochastic fluctuation at all three positions in the electrode, therefore no quantitative deviations of composition along x axis is given for this low C-rate. Structures show a SEI_1 and SEI_2 rich region close to the electrode surface and region with all three components closer to the electrolyte. In particular close to the solid and close to the electrolyte, larger clusters of the SEI_1 component can be found. The formation of clusters can be assigned to the assumption of strong surface binding of LiS_1 to SEI_1, as well as strong binding energies between neighboring SEI_1 components. Small clusters, i.e. approximately 10 molecules, of SEI_2 and SEI_3 components can be observed, as they are not randomly distributed, which is assigned to the repulsive force assumed between those components.

In Figure 4.9, the composition of the surface film as a function of the film height is shown in detail. The fraction of the three components is shown for the instance

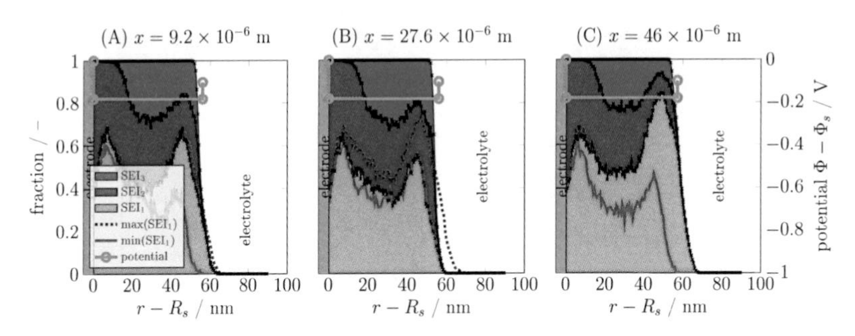

Figure 4.9: Film composition of one kMC instance after formation with 0.07C at $t = 27560$ s. Maximum and minimum values of SEI_1 fraction out of four kMC instances are indicated. Further, electrode potential at electrode (s), left and right of film, and electrolyte (e) is shown. Reproduced from publication [1] with kind permission from Wiley-VCH Verlag GmbH & Co. KGaA.

to the left of Figure 4.8 (A–C). Further, the maximum and minimum fraction of SEI_1, out of the four instances shown in Figure 4.8, is indicated to illustrate the effect of stochasticity. It can be seen that SEI_3 appears only after a certain height, $r - R_s > 10$ nm. This observation can be assigned to the Gibbs free energy of the SEI_3 decomposition, which is at lower potentials, i.e. 0.6 V, as shown in Figure 4.6. Further, there is a peak for the fraction of SEI_1 at low heights, $r - R_s = 8$ nm and higher heights, $r - R_s = 50$ nm. In Figure 4.8 it can be seen that those peaks correspond to large clusters of SEI_1. Finally, by indicating the minimum and maximum values for SEI_1 fraction, the statistical fluctuation, as observed previously, is quantified. It can be seen that fluctuation is rather high and increases with increasing film thickness. However, the two peaks in SEI_1 fraction are statistically relevant and can be found for all four instances at all positions.

Results in Figure 4.8 and 4.9 show that the chosen heterogeneity and parameter assumptions yield a distinct structure of the surface film.

In Figure 4.6, electrical potential of anode against a lithium metal reference is shown for simulation of formation with 1C and 2C, respectively. It can be seen that potentials of the anode are shifted towards lower potentials with increasing C-rates due to significant kinetic potential losses. Further, results show a slight SEI formation plateau for both simulations. The potential shifts to more negative potentials, as can be also seen in Figure 4.6.

In Figure 4.7 (B), average film thickness at the anode for formation with 1C and 2C is shown at the three positions. It can be seen that for all simulations, film thickness is continuously increasing at all positions. In general, film thickness is highest close to the separator, i.e. at position $x = 46 \times 10^{-6}$ m. For simulation of formation with 2C, the film thickness close to the separator is more than twice as high as film thickness close to the current collector. The difference in film thickness considerably increases with increasing C-rates. In simulation results shown here, film growth rate increases with increasing C-rate. For formation with 2C, the slope of the film thickness considerably differs by approximately 10% for different positions. In contrast, at formation with 1C, the slope of film thickness is comparable at all investigated positions. Further, thickness of four single instances of the kMC is shown. It can be seen that there is no significant difference between the instances and the average value. Only for 2C and small values for x, a slight fluctation of thickness can be observed. Results suggest that with higher formation rates, macroscopic differences in film thickness along electrode position may be observed. Differences in local conditions, such as species concentrations and potentials, affect atomistic processes and structuring of the film, which is discussed below. In the following, composition and film structure along electrode position is discussed in detail.

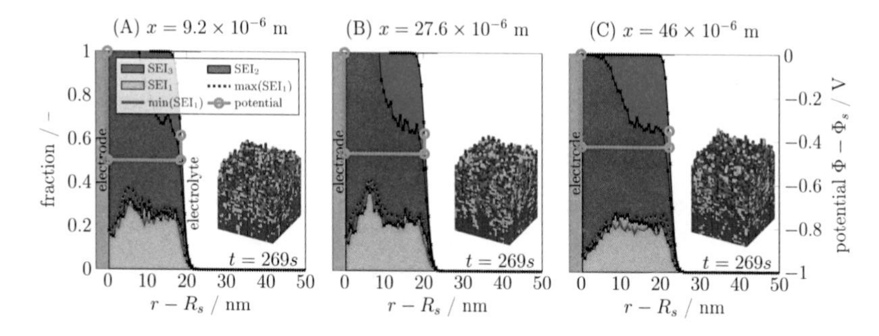

Figure 4.10: Composition and film structure of a kMC instance at $t = 269$ s during the first charge with 1C at position $x = 9.2 \times 10^{-6}$ m (A), $x = 27.6 \times 10^{-6}$ m (B), and $x = 46 \times 10^{-6}$ m (C). For SEI_1 composition, also the maximum and minimum value out of four instances is shown. Further, electrode potential at electrode (s), left and right of film, and electrolyte (e) is shown. Reproduced from publication [1] with kind permission from Wiley-VCH Verlag GmbH & Co. KGaA.

In Figure 4.10 (A–C), surface film composition along film thickness at the three electrode coordinates x is shown for 1C at 269s. This time is still during the active buildup of SEI, as seen in Figure 4.7, but it enables direct comparison with 2C formation, where 269s is at the end of formation. It is furthermore ca 100x shorter than at time for structures illustrated for 0.07C. In any case, the structures in the film remain constant with time, as modifications just occur at the surface, enabling a sound comparison at all times. First, the difference between film composition and structure between formation with 1C and 0.07C, shown in Figure 4.8 and 4.9, is discussed. As can be seen, the fraction of SEI_1 is significantly larger at lower charging rates. Further, a distinct peak of SEI_1 fraction at 10 and 45 nm, as seen with low C-rates, cannot be seen at 1C formation, but only a slight peak at 8 nm can be seen. Moreover, the clusters of SEI_1 are much smaller at higher charging rates. No significant difference in composition and structure can be seen between 0 and 10 nm. At thicknesses higher than 10 nm, a increasing content of SEI_3 can be seen. This second part is slightly thicker close to the separator. As can be seen, the statistical fluctation of SEI_1 component is significantly reduced at 1C compared to 0.07C, which denotes less local heterogeneity. Further, the surface is flatter compared to 0.07C. To conclude, increasing the charging rate yields a much more homogeneous film structure throughout the electrode compared to the slow film formation shown in the previous section, because stochastic fluctuation could be reduced, while spacial differences remain moderate.

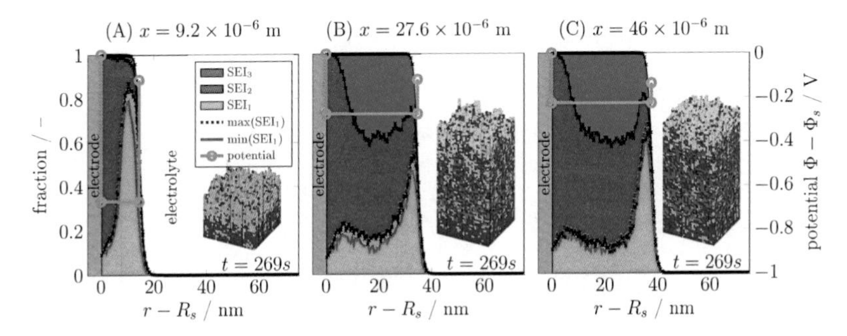

Figure 4.11: Composition and film structure of a kMC instance at $t = 269$ s during the first charge with 2C at position $x = 9.2 \times 10^{-6}$ m (A), $x = 27.6 \times 10^{-6}$ m (B), and $x = 46 \times 10^{-6}$ m (C). For SEI_1 composition, also the maximum and minimum value out of four instances is shown. Further, electrode potential at electrode (s), left and right of film, and electrolyte (e) is shown. Reproduced from publication [1] with kind permission from Wiley-VCH Verlag GmbH & Co. KGaA.

The dependency of electrode position is more distinct with higher C-rates, which is shown in Figure 4.11 for the time as for 1C formation. Here, close to the current collector, i.e. Figure 4.11 (A), SEI_3 components can barely be found. In contrast, at the other two positions, a SEI_3 rich region can be found at thicknesses larger than about 8 nm. Compared to 1C formation, the fraction of SEI_1 is lower close to the separator and higher close to the current collector. Further, at 2C discharge a peak of SEI_1 can be seen close to the surface. This peak is most distinct close to the current collector. Concerning statistical fluctuation of SEI_1 components, the trend shown with 1C continues and yields statistically reproducible structures. Further, it can be seen that statistical fluctuations are higher close to the current collector. Moreover, significant deviations in thickness, composition and structure along the x coordinate can be seen.

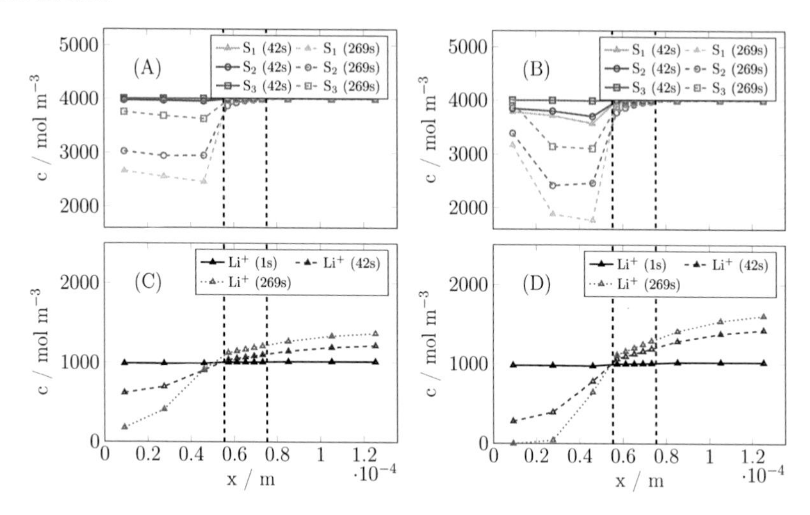

Figure 4.12: Spacial and temporal changes of concentration in the electrolyte for S_1, S_2, and S_3 (A) and concentration of Li^+ (C) for formation with 1C. Concentration in the electrolyte for S_1, S_2, and S_3 (B) and concentration of Li^+ (D) at for formation with 2C. Reproduced from publication [1] with kind permission from Wiley-VCH Verlag GmbH & Co. KGaA.

The observed dependency of composition on position x within the electrode can be assigned to macroscopic transport processes. In Figure 4.12, the concentration of lithium and solvent components along the x coordinate is shown for 1C and 2C formation, respectively. Concentrations are shown at $t = 42$ s and $t = 269$ s. Comparing Figure 4.12 (A) and (B), it can be seen that with higher C-rates, solvent components are consumed significantly faster, which leads to lower concentrations and higher gradients along the x coordinate. This can be observed likewise for the concentration of lithium-ions, which is shown in Figure 4.12 (C) and (D). It is noted that lithium-ions are consumed by both, side reaction and lithium-intercalation. In Figure 4.10 and 4.11, electrical potential from solid to the electrolyte is shown for various positions for 1C and 2C formation, respectively. Similar as has been observed with concentration gradients, potential gradients along the x coordinate increase with increasing C-rate. Comparing the trend in electrical potential along x with the trend of film thickness reveals a strong correlation between both variables. Thereby, the significant difference in potential at 2C can be assigned to high overpotentials at the electrolyte adsorption site due to depletion of lithium in the electrolyte. As the macroscopic states, such as concentration and electrical potential, depend on x, and gradients increase with increasing C-rates, the input parameters from the kMC simulations also depend on x directions, which leads to the dependencies of composition and structure shown above.

In summary, the main observations for increasing the C-rate during film formation are as follows: (I) Fraction of SEI_1 components as well as the size of SEI_1 clusters decrease with increasing film growth rates; (II) Film structure is more homogeneous; (III) Dependency of film composition, structure, and thickness on the x coordinate increases. These findings indicate interaction throughout the scales, i.e. between the electrode and the atomistic scales. Based on the results shown in this section, in the following section the major aspects of the underlying processes and their multiscale interactions are outlined.

4.5.2 Discussion of the multiscale interaction

Increasing the C-rate enforces the increasing film growth rate, because more electrons need to be transferred at the interface. With this, mechanism 1 is disadvantaged compared to mechanism 2 and mechanism 3, which is explained in the following. With faster formation procedures, growing clusters of SEI_1 are rapidly covered by the other decomposition products. This denotes that there is not enough time for the SEI_1 clusters to establish, even though they are thermodynamically favored. A single microscopic process, which covers a SEI_1 component on the surface at the beginning of the film formation, may impede the growth of large clusters at this position and thus can significantly influence the development of the film locally. This explains the significant reduction of the SEI_1 fraction with increasing film growth rate. Moreover, this also explains the high heterogeneity and stochastic fluctuations with slow formation procedures.

If charging rates are very high, as for the simulation of a 2C charge, concentrations and electrical potential depend significantly on distance to the current collector. Depletion of lithium-ions close to the current collector leads to very low ionic conductivity and high overpotentials for lithium adsorption. Therefore, at this position the reaction rates at the surface are significantly reduced and yield a slower film growth rate. As slow growth favors SEI_1, this leads to gradient in fractions of SEI_1 along the x coordinates, higher fractions close to the current collector, due to slow film growth rate, and lower fractions close to the separator, due to high film growth rates. Moreover, as film growth slows down, SEI_1 clusters are formed on the surface at all positions. To conclude, with fast charging procedures, a significant difference in film structure and composition along the x coordinate can be observed.

The effects of the interaction observed here are strongly related to the mechanism and parameters, which have been chosen. However, it can be expected that a comparable interaction can be found with the actual decomposition mechanism, which is much more complex and most certainly heterogeneous as well, as batteries nowadays use a multicomponent mixture of electrolytes which decompose differently.

Ageing of batteries is known to be caused by local heterogeneity [132], thus the optimization of the film structure should aim for homogeneous film throughout the electrode. The presented results indicate that local heterogeneity could be caused by both, too slow charging protocols, as stochastic effects prevail, and too fast charging protocols, causing spacial distributions. These findings suggest that there is an optimal charging rate for a certain electrolyte composition, and that todays state-of-the-art slow formation protocols cannot necessarily be considered optimal for long life time. Another optimization objective could be to achieve certain optimal properties of the SEI, e.g. mechanical stability or high ionic conductivity. These are possibly related to the cluster size of certain SEI components, such as for the SEI_1 component of the example provided here. The presented results indicate that cluster size can be controlled by the charging procedure. Both suggested optimization tasks require consideration of the multiscale effects, as revealed in this chapter.

4.6 Concluding remarks

In this chapter the P2D model is extended by a kMC model of surface film growth, analogous to chapter 3. Main parameters of the model were exemplary identified by experimental data from electrochemical measurements. It could be seen that the model can describe the potential slope during the formation process. Further, the approach shows how a consistent model formulation and a step wise parameter identification can be applied. The model was than further used to illustrate multiscale effects, which can occur during battery cycling with low and high c-rates. Results indicated evidence for multiscale effects due to spacial distribution concentration in the electrode. Further, it could be seen that different growth rates as result of applied current, can impact the film and electrode structure, while at low charging rates a strong impact of stochastic effects leads to a much more heterogeneous films.

To conclude, this chapter presented a methodology to analyze multiscale effects in the formation process and thereby provided simulation results, which strongly motivates to consider such effects in future.

Chapter 5

Macroscopic Heterogeneity[6]

5.1 Introduction

As has been outlined in chapter 1, the macroscopic structure has a significant impact on the atomistic processes at electrochemical active surfaces. Macroscopic electrochemical models are widely used to investigate lithium-ion batteries with the aim to predict their electrochemical performance. It is usually neglected that the electrodes are highly heterogeneous. Heterogeneity can be, for instance, an uneven electrode thickness, a pore size distributions, a heterogeneous distribution of components, or a particle size distribution (PSD). All have one thing in common, namely that they cause heterogeneous local conditions, e.g. concentrations, potentials, or current densities. As shown in previous chapters, i.e. chapter 3 and 4, these local conditions impact degradation processes like surface film growth and thus may need to be considered in detail. Therefore, in this chapter the effect of heterogeneity on local conditions and performance as well as the degradation on electrode scale is investigated at the example of PSD.

The PSD of the active material is a well known property in lithium-ion batteries and can be adjusted during the manufacturing of battery electrodes [146, 147]. It is presumed that the particle size itself, as well as its distribution, affects the capacity and ageing behavior of lithium-ion batteries significantly [146], which was also analyzed in chapter 3. However, the relationship of battery performance and degradation with particle size distribution is currently barely addressed in research. Hence, this chapter aims to use a mathematical model that combines the electrochemical features of a battery electrode and the impact of the particle size distribution of the active material. Simulation results will be shown for the example of graphite electrodes. In the first part in section 5.2, the effect of particle size distribution on performance of graphite electrodes is investigated. This reveals conditions where an accurate consid-

[6]Part of this chapter has been published in (Röder et al., Energy Technology, 4, 1588-1597, 2016 [60])

eration of particle size distribution is needed to reproduce battery performance, and where homogeneous approaches are feasible for this purpose. Further, the impact of particle size on local current densities is analyzed. In the second part in section 5.3, the effect of change of electrode structure due is investigated with a scenario-based approach.

5.2 Comparing homogeneous and heterogeneous models at the example of particle size distribution

In this section, the effect of particle size distribution is investigated. Whereas it is well known that smaller sized particles possess better performance, the impact of the actual distribution shape and scale is often not investigated. Thus, it is of particular interest to further elucidate how the various particle sizes influence the electrochemical performance. Furthermore, several battery degradation effects are linked to high charging rates. This causes high electrochemical potentials and high intercalation rates [7], which can differ locally with particle size in a distributed heterogeneous system. Since the volume to surface ratio varies with size, the intercalation rate of lithium through the surface of a host structure is linked to the particle size. The degradation progress might differ with the particle size of the active material in such systems [144].

There have been several publications in the last decades which show the impact of the mean particle size on battery performance [148, 149]. Farkhondeh et al. present an electrochemical model with a PSD of the active material with three different sized particle groups [62]. The PSD is thereby determined by scanning electron microscopy. They show that neglecting the distribution of particles within the electrode largely underestimates the capacity at the end of a discharge with elevated C-rates. However, single particle model approaches are limited as they do not accurately describe the battery performance at higher current densities, so Farkondeh et al. [63] also present a porous electrode model to investigate the impact of C-rates greater than 1 C. Their work demonstrates that the effect of PSD can be modeled with such an approach.

In this section, a graphite electrode of a lithium-ion battery is described mathematically with distributed particle sizes of the active material, which are intentionally adjusted to a certain PSD. The general impact of shape and size of this distribution on the electrode performance is investigated. Obtained results are compared with the single particle model, as used in chapter 3, to evaluate under which conditions the PSD has to be taken into account. Furthermore, it is shown to what extent the PSD affects the current density at the surface of the differently sized particles, since the local current density is a key property for several degradation effects, such as the SEI formation.

5.2.1 Mathematical models

Equations of the mathematical model are based on the commonly used single particle approach as well as its modification by consideration of multiple particles with different size. Both modeling approaches contain simplifications to single out the effect of diffusion in electrodes with different PSDs. Required assumptions and the consequential validity are shown in Table 5.1.

Table 5.1: Model assumptions and consequential validity.

No	Assumption	Validity
1	electrolyte and electrical resistance is neglected	low C-rates
2	electrical potential at particle is independent of position within the electrode	low C-Rates / good wetting and homogeneous active material fraction
3	efficient solid diffusion coefficient independent of particle size	homogeneous particle structure
4	particles are ideal spheres	sphere like particles
5	particle and material properties independent of PSD	manufacturing of different PSD does not impact particle properties (e.g efficient diffusion and activity coefficients)

Homogeneous Single Particle Model

An electrode model is introduced based on the widely used single particle model approach [45]. This model includes a charge balance equation and solid diffusion equation within the particles. The diffusion equation is written as:

$$\frac{\partial c_{\mathrm{Li(s)}}(r)}{\partial t} = \frac{1}{r^2}\nabla(D_s^{\mathrm{Li(s)}} r^2 \nabla c_s^{\mathrm{Li(s)}}(r)) \tag{5.1}$$

with the boundary conditions $-D_s^{\mathrm{Li(s)}}\nabla c_s^{\mathrm{Li(s)}}(r=0) = 0$ and $-D_s^{\mathrm{Li(s)}}\nabla c_s^{\mathrm{Li(s)}}(r =$

$R_s) = {}^{J^{\mathrm{Li}}}/_{\mathrm{F}}$, where Faraday constant F, lithium reaction current density $J^{\mathrm{Li}} = {}^{j^{\mathrm{Li}}}/_{a_s}$, specific surface area $a_s = {}^{3\varepsilon_s}/_{R_s}$, active material fraction ε_s and particle radius R_s. The charge balance is considered at the electrochemical double layer with [24, 134]:

$$a_s C^{\mathrm{DL}} \frac{\partial \Delta\Phi}{\partial t} = j^{\mathrm{charge}} - j^{\mathrm{Li}} \tag{5.2}$$

with double layer capacitance C^{DL} and electrode potential $\Delta\Phi$. The intercalation reaction at the solid particle / electrolyte (s/e) interface is:

$$\mathrm{Li(s)} \rightleftharpoons \mathrm{Li^+(e)} + \mathrm{V(s)} + \mathrm{e^-(s)} \tag{5.3}$$

with lithium in the solid Li(s), vacancies in the solid V(s) electrons in the solid $\mathrm{e^-(s)}$, and lithium-Ions in the electrolyte $\mathrm{Li^+(e)}$. The surface current density J^{Li} through this lithium reaction can be determined by:

$$\begin{aligned} \frac{J^{\mathrm{Li}}}{F} &= C^0_{\mathrm{Li(s)}} a^{\mathrm{Li(s)}} k_a \exp\left(\frac{\beta \Delta\Phi F}{RT}\right) \\ &\quad - C^0_{\mathrm{Li^+(e)}} a_{\mathrm{Li^+(e)}} C^0_{\mathrm{V(s)}} a^{\mathrm{V(s)}} k_c \exp\left(\frac{-(1-\beta)\Delta\Phi F}{RT}\right) \end{aligned} \tag{5.4}$$

where the activity coefficients a^k are either assumed to be ideal $a^k = {}^{c^k}/_{C^0_k}$ (e.g. $\mathrm{Li^+(e)}$) or determined with a semi empirical function (e.g. for Li(s) and V(s)), while C^0_k is the standard state concentration of species k. The relation between anodic and cathodic reaction rate constant is [117]:

$$\frac{k_a}{k_c} = \frac{C^0_{\mathrm{Li(e)}} C^0_{\mathrm{V(s)}}}{C^0_{\mathrm{Li(s)}}} \exp\left(-\frac{\Delta G^0}{RT}\right) \tag{5.5}$$

with change in Gibbs free energy $\Delta G^0 = \mu^0_{\mathrm{Li(e)}} + \mu^0_{\mathrm{V(s)}} - \mu^0_{\mathrm{Li(s)}}$, where μ^0_k is the standard chemical potential of species k.

Heterogeneous Multiple Particle Model

For multiple particle modeling, the presented model is further extended by the effect of PSD [61, 40, 63, 62]. Using the framework provided in [150], equations are derived for the distributed parameters. Number density, surface area density, and volume density are determined based on the number fraction density. Assuming spherical particles,

$A_s(R_s) = 4\pi R_s^2$ and $V_s(R_s) = {}^4/_3\pi R_s^3$ define the surface area and the volume of a particle radius R_s, respectively. This is used to determine the surface area density and the volume density based on the number density:

$$f_{\mathrm{vol}}(R_s) = f_{\mathrm{num}}(R_s)V_s(R_s) \tag{5.6}$$

$$f_{\mathrm{area}}(R_s) = f_{\mathrm{num}}(R_s)A_s(R_s) \tag{5.7}$$

In general, the integral over these number density functions is equal to the total surface area ratio, a_{s}, and the solid phase volume fraction, ε_{s}, respectively, so that:

$$\varepsilon_{\mathrm{s}} = \int_0^\infty f_{\mathrm{vol}}(R_s)\mathrm{d}R_s \tag{5.8}$$

$$a_{\mathrm{s}} = \int_0^\infty f_{\mathrm{area}}(R_s)\mathrm{d}R_s \tag{5.9}$$

The relation between number fraction density and number density is given by the total number of particles per volume n_s:

$$f_{\mathrm{num}}(R_s) = n_s h_{\mathrm{num}}(R_s) \tag{5.10}$$

Combining equation (5.10) with equation (5.6) and (5.9) yields:

$$\varepsilon_{\mathrm{s}} = n_s \int_0^\infty h_{\mathrm{num}} V_s(R_s)\mathrm{d}R_s \tag{5.11}$$

With ε_{s}, n_s is eliminated, the relation between number fraction density and number density is introduced as:

$$f_{\mathrm{num}}(R_s) = \frac{\varepsilon_{\mathrm{s}} h_{\mathrm{num}}(R_s)}{\int_0^\infty h_{\mathrm{num}}(R_s)V_s(R_s)\mathrm{d}R_s} \tag{5.12}$$

The surface area fraction density, $h_{\mathrm{area}}(R_s)$, and the volume fraction density, $h_{\mathrm{vol}}(R_s)$, are of particular interest for analysis and understanding the electrochemical system. They are defined as follows:

$$h_{\mathrm{vol}}(R_s) = \frac{f_{\mathrm{num}}(R_s)R_s^3}{\int_0^\infty f_{\mathrm{num}}(R_s)R_s^3\mathrm{d}R_s} \tag{5.13}$$

$$h_{\mathrm{area}}(R_s) = \frac{f_{\mathrm{num}}(R_s)R_s^2}{\int_0^\infty f_{\mathrm{num}}(R_s)R_s^2\mathrm{d}R_s} \tag{5.14}$$

The mean values $\bar{R}_s^{\mathrm{num}}$, $\bar{R}_s^{\mathrm{area}}$ and $\bar{R}_s^{\mathrm{vol}}$ of the fraction density functions are determined using:

$$\bar{R}_s = \int_0^{\infty} R_s h(R_s)\mathrm{d}R_s \tag{5.15}$$

To extend the presented single particle model by the effect of diffusion in multiple particles the diffusion equation (5.1) is considered for every particle with the radius R_s as:

$$\frac{\partial c_{\mathrm{Li}_s}(r, R_s)}{\partial t} = \frac{1}{r^2}\frac{\partial}{\partial r}\left(D_s r^2 \frac{\partial c_{\mathrm{Li}_s}(r, R_s)}{\partial r}\right). \tag{5.16}$$

The electrode potential at every particle of the size R_s is the same (compare model assumptions Table 5.1), so reaction overpotential and reaction rate only depend on the local activities, which themselves are functions of the surface concentration at the particle surface. Concentration values are particularly affected by state of charge (SOC) and diffusion within a single particle. This results in the modified definition of boundary conditions for different particles as $-D_s \partial c^i_{\mathrm{Li}_s}(r{=}R_s, R_s)/\partial r = J^{\mathrm{Li}}(R_s)/F$. The reaction density j^{Li} is finally determined by:

$$j^{\mathrm{Li}} = \int_0^{\infty} f_{\mathrm{area}}(R) J^{\mathrm{Li}}(R)\mathrm{d}R \tag{5.17}$$

For a numerical solution, equations are discretized in finite volumes in r and R_s domain.

5.2.2 Model parameters

To give a fundamental and general view on performance and degradation of graphite electrodes, system parameters and simulations are described in details in this section.

Kinetic and Thermoynamic Parameters

Several material properties are needed for model parametrization. The diffusion coefficient of graphite materials commonly used in simulations ranges orders of magnitude from $2 \cdot 10^{-16}$ - $5 \cdot 10^{-9}$ $\mathrm{m^2 s^{-1}}$ [151, 134, 117, 152, 153]. Experimental results for MCMB particles suggest diffusion coefficients of $1 \cdot 10^{-15}$ - $1 \cdot 10^{-13}$ $\mathrm{m^2 s^{-1}}$ [154]. Good agreement to discharge curves by simulations is often obtained by rather small diffusion coefficients. Furthermore the effect of PSD is more distinct with smaller diffusion coefficient. Therefore a diffusion coefficient is chosen at the lower bound of the experimentally determined coefficients with $1 \cdot 10^{-15}$ m^{2}s^{-1}. Nevertheless, it needs to be emphasized that presented effects of PSD on performance of graphite electrodes may be less distinct for graphite materials with faster solid diffusion.

Standard chemical potentials for lithium, vacancies, and electrons are provided by Colclasure et al. [117]. Standard concentration of lithium and vacancies in graphite are defined to their maximum concentration c_{max} and for lithium in the electrolyte to 1200 mol m^{-3}. The non ideality of lithium and vacancy activity in graphite is provided by Redlich Kister coefficients [117]. The rate of the ideal concentration dependent exchange current density is given in [117] with $k_{ct} = 1.429 \cdot 10^{-9}$ m$^{2.5}$ mol$^{-0.5}$ s^{-1} based on experimental data of [151]. Ideal exchange current density assumes ideal activity $a_i = {}^{c_i}/_{C_i^0}$ and can be written as:

$$i_0^{ideal} = k_{ct} F (c^{\mathrm{Li^+(e)}})^{\beta} (c_{max} - c^{\mathrm{Li(s)}})^{\beta} (c_{\mathrm{Li(s)}})^{(1-\beta)} \tag{5.18}$$

Non-ideal exchange current density, as applied in this chapter, includes non ideality of activity through Equation 5.4 and can be written as:

$$i_0 = c_{max} F k_a \exp\left(\frac{\Delta G^0 \beta}{RT}\right) \left(a^{\mathrm{Li^+(e)}}\right)^{\beta} \left(a^{\mathrm{V(s)}}\right)^{\beta} \left(a^{\mathrm{Li(s)}}\right)^{(1-\beta)}, \tag{5.19}$$

If the assumption for ideal activity coefficients $a^k = {}^{c^k}/_{C_k^0}$ is applied to equation 5.19 this leads to another representation of equation 5.18:

$$i_0^{ideal} = c_{max} F k_a \exp\left(\frac{\Delta G^0 \beta}{RT}\right) \frac{(c^{\mathrm{Li^+(e)}})^{\beta}}{(C^0_{\mathrm{Li^+(e)}})^{\beta}} \frac{(c_{max} - c^{\mathrm{Li(s)}})^{\beta}}{(C^0_{\mathrm{V(s)}})^{\beta}} \frac{(c^{\mathrm{Li(s)}})^{(1-\beta)}}{(C^0_{\mathrm{Li(s)}})^{(1-\beta)}} \tag{5.20}$$

and thus the given parameter for reaction rate constant k_{ct} can be used to determine the rate constant k_a of the non-ideal exchange current density:

$$k_a = k_{ct} \frac{(C^0_{\mathrm{Li(s)}})^{(1-\beta)} (C^0_{\mathrm{V(s)}})^{\beta} (C^0_{\mathrm{Li^+(e)}})^{\beta}}{c_{max}} \exp\left(\frac{-\Delta G^0 \beta}{RT}\right) \tag{5.21}$$

All parameters used in this chapter are provided in the appendix A in Table A.10.

Particle Size Distribution

The effect of PSD on the electrode structure is investigated within the usual range for graphite materials. To compare different PSD the theoretical capacity and thus the amount of active material needs to be fixed. Therefore active material fraction is chosen with $\varepsilon_s = 0.6$ as a PSD independent parameter.

A good indicator for the property of a PSD is the R_{50} value and the R_{50}/R_{90}

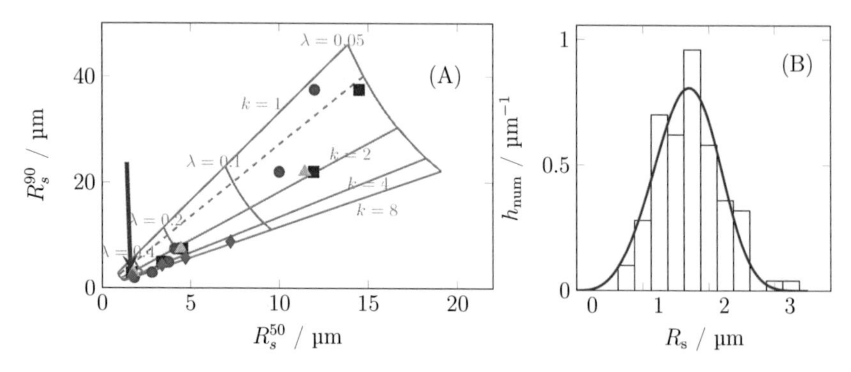

Figure 5.1: (A) Typical vales for R_s^{50} and R_s^{90} for different graphites: E-LSG (▲) [156], SFG (■) [156], KS (●) [156] and MCMB (◆) [155] including corresponding values of Weibull distributions in this range. (B) Particle Size Distribution of a MCMB taken from [155] including a fit of a Weibull distribution (-). Reprinted from publication [60].

ratio. Figure 5.1 (A) shows R_{50} vs. R_{90} for different graphite materials as given in Literature [155, 156]. The arrow indicates a PSD of MCMB given by Baohua et al. [155], which is is shown in Figure 5.1 (B) including a fit by application of Weibull distribution:

$$h_{\text{num}}(R_s) = \lambda \cdot k \cdot (\lambda \cdot (R_s))^{k-1} \cdot \exp\left(-(\lambda \cdot (R_s))^k\right) \tag{5.22}$$

The Figure shows in addition the correlation between λ and k and their influence on R_s^{50} and R_s^{90}. In this work, performance of graphite electrodes is investigated for different theoretical Weibull distributions within the presented typical range of PSD.

5.2.3 Discussion of the impact of macroscopic heterogeneity

The presented models enable one to study the electrode potential during the discharge process as shown in Figure 5.2 for a graphite electrode. The electrode potential of a graphite electrode vs. a lithium reference, as simulated here, increases during discharge. The maximum electrode capacity is defined at a cut-off voltage of 1 V. Higher electrode potentials are usually evoked by higher internal reaction or diffusion resistances and imply a worse electrode performance, as well as a lower maximum discharge electrode capacity. Maximum theoretical discharge capacity C_{Ah} is used to define a dimensionless electrode capacity. Reaction and diffusion resistances decrease with decreasing charging rates, as can be seen in Figure 5.2, where the dimensionless electrode capacity is almost 1 for C rates of 0.01 C, but considerably lowered for

Table 5.2: Electrode discharge capacity for 1C discharge with different shape and scale parameters of the PSD. Capacity of for the multi particle model is marked bold and capacity with deviations large than 1% are marked red and underlined. Reprinted from publication [60].

shape	model approach	$\lambda = 0.8$	$\lambda = 0.4$	$\lambda = 0.2$	$\lambda = 0.1$	$\lambda = 0.05$
$k = 8$	PSD	**0.968**	**0.878**	**0.58**	**0.222**	**0.072**
	R_s^{num} approx.	0.972	0.893	0.612	0.239	0.077
	R_s^{area} approx.	0.97	0.884	0.586	0.222	0.072
	R_s^{vol} approx.	0.969	0.879	0.575	0.216	0.07
$k = 4$	PSD	**0.961**	**0.853**	**0.537**	**0.201**	**0.067**
	R_s^{num} approx.	0.974	0.901	0.635	0.254	0.081
	R_s^{area} approx.	0.966	0.87	0.552	0.203	0.067
	R_s^{vol} approx.	0.963	0.857	0.522	0.187	0.063
$k = 2$	PSD	**0.924**	**0.736**	**0.385**	**0.132**	**0.052**
	R_s^{num} approx.	0.975	0.905	0.648	0.264	0.084
	R_s^{area} approx.	0.946	0.789	0.404	0.133	0.051
	R_s^{vol} approx.	0.931	0.734	0.338	0.108	0.046
$k = 1.5$	PSD	**0.868**	**0.61**	**0.272**	**0.091**	**0.044**
	R_s^{num} approx.	0.974	0.902	0.637	0.256	0.081
	R_s^{area} approx.	0.914	0.678	0.287	0.091	0.043
	R_s^{vol} approx.	0.88	0.575	0.216	0.07	0.04

higher rates of 1 C.

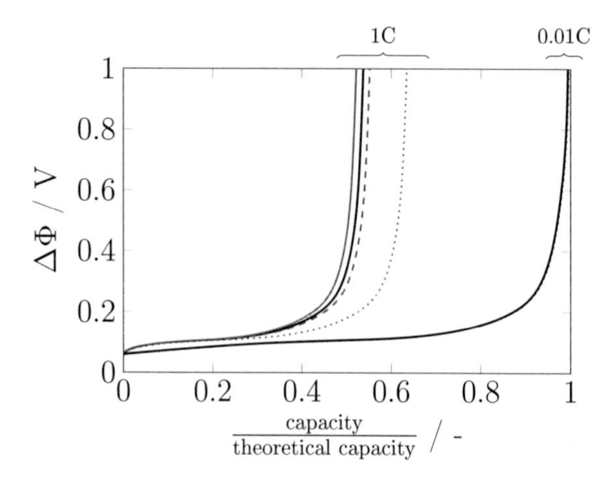

Figure 5.2: Electrode potential during discharge process for different C-rates applied, shown for accurate consideration of the PSD using a multiple particle model (—) and its approximation using a single particle model based on number (······) , surface area (- -) and volume based (—) mean radii. Reprinted from publication [60].

The PSD scale of a Weibull distribution is defined by the parameter λ. Small λ values denote large scaled distributions, i.e. large distribution width and large mean radius. Figure 5.3 shows the impact of the distribution scale. It can be seen that with increasing width of the PSD, the electrode capacity decreases. Additionally, the potential of the plateau is higher with smaller λ values, which can also be confirmed by experimental results [156]. Both effects are caused by higher internal resistances through longer diffusion pathways in larger particles and smaller specific surface area a_s in distributions with larger particles. The impact of PSD shape on the electrode capacity is shown in Figure 5.4. This figure shows the change in electrode capacity is dependent on shape parameter k for different PSD scales λ. It can be seen that the shape parameter k decreases, i.e. increasing width of PSD, while maintaining almost constant number based mean radius. In addition, the maximum electrode capacity decreases, while the gradient in this case is larger for smaller k values. The decrease in maximum electrode capacity is caused by an increasing number of larger particles having longer diffusion pathways and leading to a lower specific surface area of the PSD. Electrode capacities for all PSDs investigated with a C-rate of 1 C are provided in Table 5.2. The best electrode performance is obtained for PSDs with large k and large λ values, i.e. narrow distributions with small mean particle radii. This denotes that it may is beneficial to tune electrode particle size by reducing amount of large sized particles.

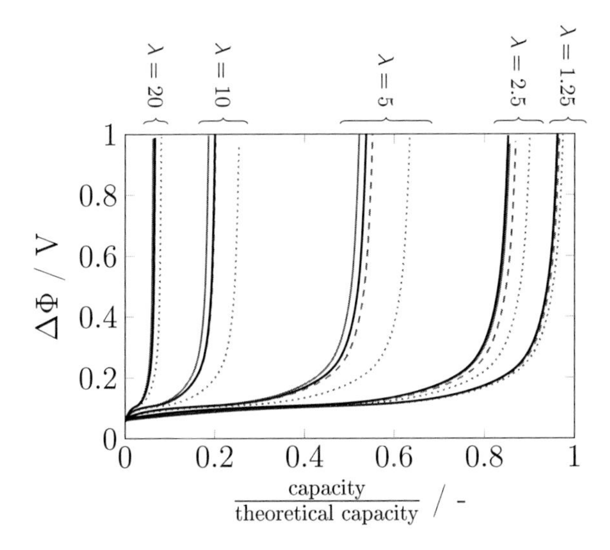

Figure 5.3: Electrode potential during discharge process for differently scaled PSDs, shown for accurate consideration of the PSD using a multiple particle model (—) and its approximation using a single particle model based on number (······) , surface area (- -) and volume based (—) mean radii. Reprinted from publication [60].

Since an accurate consideration of PSD using a multiple particle model is computationally much more expensive than a single particle model, often an approximation using a single particle model is preferable. Different approaches to determine a mean radius are possible in general based on number, surface area or volume distributions. To evaluate how far they can accurately represent the multi particle behavior, previously introduced simulation are compared to results obtained by the multiple particle model with results generated by using a single particle model with the different approximations of the mean particle radius, i.e. $\bar{R}_s^{\text{num}}$, $\bar{R}_s^{\text{area}}$ and $\bar{R}_s^{\text{vol}}$. Figure 5.2 shows that curves for low C-rates are all overlaying. Due to very low kinetic losses, there is no considerable deviation between both models and different approximation. In contrast, at higher C-rates simulation results considerably differ. Surface area and volume based mean approximation slightly overestimate and underestimate electrode potential, respectively. In contrast, the electrode potential is clearly underestimated by the number based mean approximation leading to a significantly overestimated maximum capacity. The results, shown in Figure 5.3, indicate that the observed

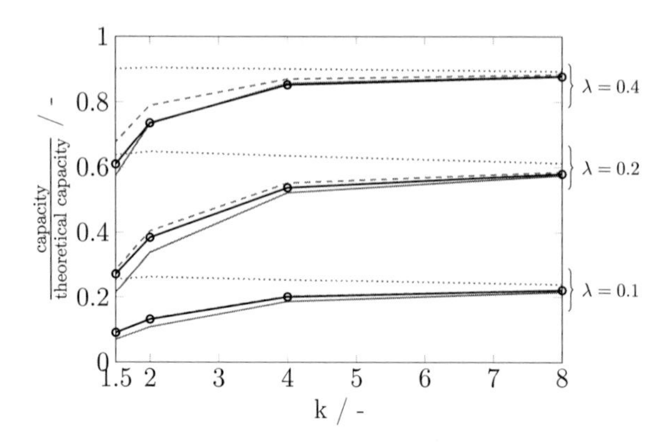

Figure 5.4: Electrode discharge capacity dependent on PSD shape, shown for differently scaled PSDs with accurate consideration of the PSD using a multiple particle model (—) and its approximation using a single particle model based on number (······) , surface area (- -) and volume based (—) mean radii. Reprinted from publication [60].

deviations are less distinct with small and large scaled PSDs than for intermediate values of λ. Furthermore, particularly in Figure 5.3, it can be seen that for small scaled PSDs, volume based approximations accurately reproduce electrode capacity, while for large scaled PSDs, surface area based approximations almost perfectly agree to the multiple particle model. The volume based mean approximation uses the most representative particle size in terms of particle volume and therefore provides a good approximation for the diffusion pathway and diffusion induced internal resistance. The reason for the good agreement of surface area based mean $\bar{R}_s^{\text{area}}$ is due to its direct correlation to specific surface area a_s:

$$a_\mathrm{s} = \int_0^\infty f_\mathrm{area}(R_s)\mathrm{d}R_s = \frac{3\varepsilon_s}{\bar{R}_s^\mathrm{area}} \tag{5.23}$$

In case of a dominating kinetic losses of reaction at the surface, this is a physically meaningful approximation. Thus, the results as such indicate that internal resistance at large scale distribution is dominated by high resistance of the surface reactions through low specific surface area. In contrast in small scaled PSDs, diffusion limitations dominate.

Furthermore, in Figure 5.4 it can be seen that both approximations are useful to qualitatively predict the impact of PSD shape in contrast to a number based mean approximation. The number based mean approximation only for very large k values, i.e. narrow PSDs, accurately predicts electrode capacity.

Table 5.2 gives an overview for electrode capacities simulated with accurate consideration of PSD using the multiple particle model and the approximations using the single particle model. Red and underlined entries denote that electrode capacity deviates more than 1% from the respective result obtained by accurate consideration of PSD, which means all black entries are in good agreement. Number based mean approximations are only accurate for few cases of narrow distributions. Surface area based approximation have small deviations for large scale distributions and volume based approximations for small distributions. Single particle approximations are least accurate for intermediate scaled and coarse distributions. The presented results suggest that in case of surface reaction or solid diffusion limited electrodes, surface area or volume based mean approximations should be used, respectively. Due to change of surface area to volume ratio for differently sized particles, heterogeneity in local surface current density can be expected. Geors et al. [144] investigated the existence of such a surface current density distribution and studied its effect on the exfoliation behavior of graphite electrodes. Exfoliation is caused by solvent co-intercalation, which as the intercalation reactions depends on local overpotentials caused by deviations in surface activity, concentration of lithium and vacancies, which is mainly affected by diffusion kinetics inside a particular particle. This lead to heterogeneous passivation, i.e. SEI layer growth [157, 144]. The reason for passivation or exfoliation of some particles may be explained by deviations in current density [144]. To evaluate local current densities on particle surfaces for different PSDs, surface area fraction density h_{area} and surface current densities $J^{\text{Li}}(R_s)$ for a 1C discharge process is shown in Figure 5.5. Surface area fraction density is shown together with surface current density, because it indicates the actual fraction of a local current density on the total surface area. Current density is shown at the end of discharge for distributions with the shape parameters $k = 8$ (Figure 5.5 (A)), $k = 4$ (Figure 5.5 (B)), $k = 2$ (Figure 5.5 (c)) and $k = 1.5$ (Figure 5.5 (D)). Further, the impact of differently scaled PSDs, i.e. scale parameter $\lambda = 0.8$ and $\lambda = 0.05$ are presentded. In general it can be seen that with larger scaled PSDs, i.e. smaller λ values, current densities are larger and with coarser PSDs, i.e. smaller k values, current densities more strongly differ between differently sized particles. Focusing on the narrow distributions in Figure 5.5 (A) and (B) it can be seen that in large scaled PSDs current density is almost constant within the PSD range, while for smaller scaled PSDs current can be considerably larger for larger particles than for smaller ones. This indicates a higher impact of diffusion processes in the particle volume at smaller scaled PSDs and is supported by the previously reported good approximation through volume based means for such distributions. Even larger differences in local current density can be observed in coarser PSDs in Figure 5.5 (C) and (D), which is supported by the experiments of Buqa et al. [156], showing high extent of exfoliation in coarse PSD. The presented simulation results indicate highest current densities in coarse and large scale PSDs at the larger particles, which can also explain why a significant lower degree of exfoliation is found at the smaller particles in experiments [144].

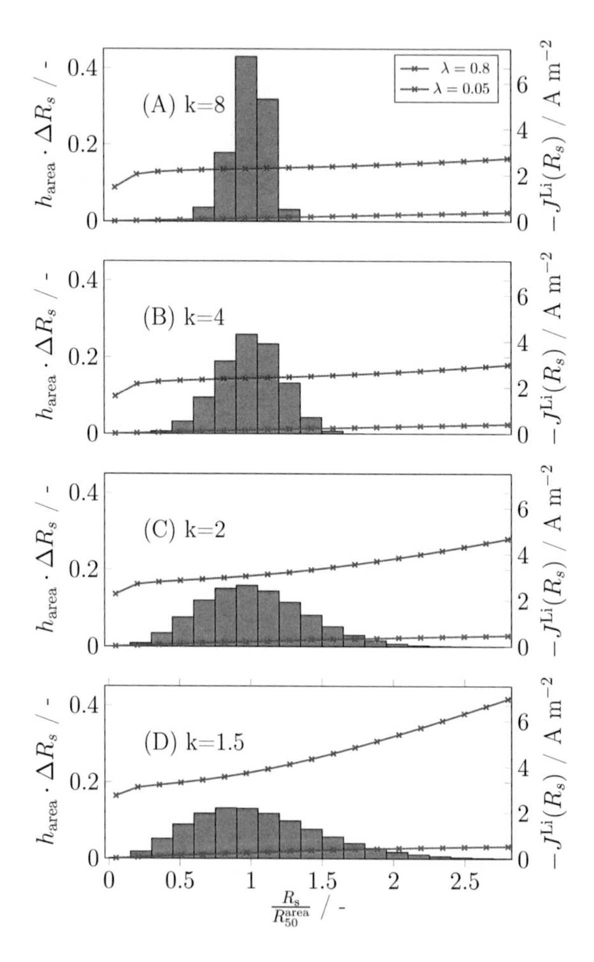

Figure 5.5: Local surface current density and surface area fraction density, shown for different particle shapes (A-D) and differently scaled PSDs at the end of discharge. Reprinted from publication [60].

In Figure 5.6 focus is laid on the coarse distribution, since observed effects are most distinct there. The change of surface current densities during the discharge process is shown for large and small scaled PSDs. The results indicate that in large scale distributions, current density is highest at the beginning of discharge and then adjusts between small and large particles, while it is contrary and less distinct for small scaled distributions. This denotes that, in particular for large scale PSDs, the actual charging/discharging can be critical in terms of high local surface current densities. Further, it can be seen that at the end of discharge, current density decreases, beginning with the smallest particles, and increases at largest particles, as long as the cut of voltage is not reached. To sum up, the presented analysis of current density with different PSDs shows in good agreement with experimental works [156, 144] that in order to avoid high current density and thus degradation processes like exfoliation, small scaled and narrow PSDs should be favored, while high current densities at the beginning of discharge should be avoided.

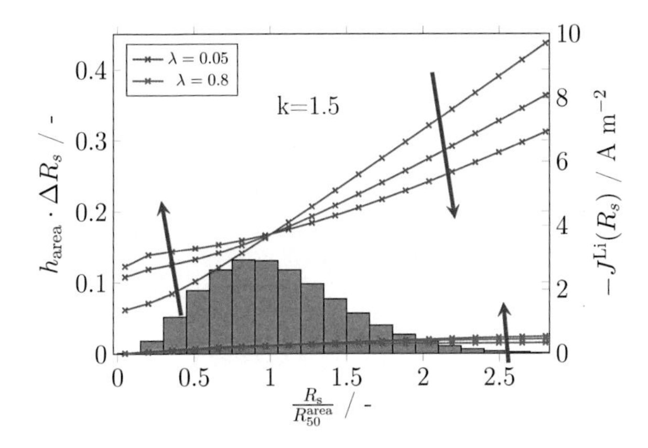

Figure 5.6: Local surface current density and surface area fraction density, shown for particle shape $k = 1.5$ and differently scaled PSDs with $\lambda = 0.8$ and $\lambda = 0.05$. Current densities are shown as changes during the discharge process for 1/3 discharge, 2/3 discharge and full discharge, while direction of change is indicated by red arrows. Reprinted from publication [60].

5.3 Degradation of the electrode structure

The general effect of PSD on performance and local conditions has been discussion in the previous section. Nevertheless, degradation of lithium-ion batteries is also related to the PSD of the active material as discussed in chapter 1. In graphite anodes, solvent co-intercalation, as well as gas evolution, can lead to particle cracking [8, 7, 158]. Additionally, the volume change of particles may cause a contact loss between particles [7]. The agglomeration of particles is observed for various materials in lithium-ion batteries [9, 10, 11], which is of particular importance for materials that undergo large volume changes [10], or that employ nano sized particles [11].

Zavalis et al. [40] uses a multiple particle model in combination with an impedance analysis to determine a possible change in the PSD during battery cycling through parameter fitting. This means that the actual evolution is not explicitly modeled. To mathematically describe the evolution of the PSD during operation, Rinaldo et al. [159] show how to apply the theory of population balances by Marchisio et al. [160] on a nano particle catalyst system. Such an approach has not be used to model the change of electrode structure in lithium-Ion batteries.

In this section populations balances are employed to analyze the general impact of change of electrode structure in a scenario based analysis [161]. With this it is shown how a change of the distribution, for example due to cracking or agglomeration of the active particles, can affect the performance of the lithium-ion battery.

5.3.1 Degradation model

Degradation on the electrode level is modeled. This includes a change of the electrode structure (i.e. change of PSD) and a change of active material fraction ε_s (i.e. electrical disconnection). Those effects can be caused by particle cracking and agglomeration through degradation processes such as solvent co-intercalation or mechanical stress through volume changes during lithiation. The structural change is modeled by applying population balances, as provided by Marchisio et al. [160]:

$$\frac{\partial f_\mathrm{num}(R_s)}{\partial t} = B^\mathrm{agl}(R_s) - D^\mathrm{agl}(R_s) + B^\mathrm{cr}(R_s) - D^\mathrm{cr}(R_s) \tag{5.24}$$

This population balance includes birth and death through aggregation and cracking as follows:

$$\begin{aligned} B^\mathrm{agl}(R_s) &= \frac{R_s^2}{2}\int_0^{R_s} \frac{\beta^\mathrm{agl}((R_s^3-P_s^3)^{\frac{1}{3}},P_s)}{(R_s^3-P_s^3)^{\frac{2}{3}}} \\ &\quad \cdot f_\mathrm{num}((R_s^3-P_s^3)^{\frac{1}{3}}) f_\mathrm{num}(P_s)\mathrm{d}P_s && (5.25) \\ D^\mathrm{agl}(R_s) &= f_\mathrm{num}(R_s)\int_0^{\infty} \beta^\mathrm{agl}(P_s,R_s) f_\mathrm{num}(P_s)\mathrm{d}P_s && (5.26) \\ B^\mathrm{cr}(R_s) &= \int_{R_s}^{\infty} \alpha^\mathrm{cr}(P_s) b^\mathrm{act}(R_s|P_s) f_\mathrm{num}(P_s)\mathrm{d}P_s && (5.27) \\ D^\mathrm{cr}(R_s) &= \alpha^\mathrm{cr}(R_s) f_\mathrm{num}(R_s) && (5.28) \end{aligned}$$

with aggregation kernel β^{agl}, cracking kernel α^{cr} and number of active fragments b^{act} with the radius R_s through cracking of a particle with the radius P_s. The resulting fragments can be either connected (i.e. active) or electrically disconnected fragments b^{dis}, which in the latter case results in a decrease of the active material fraction ε_{s} by:

$$\frac{\partial \varepsilon_{\text{s}}}{\partial t} = -\int_0^{\infty} \left[\int_{R_s}^{\infty} \alpha^{\text{cr}}(P_s) b^{\text{dis}}(R_s|P_s) f_{\text{num}}(P_s) \text{d}P_s \right] V_s(R_s) \text{d}R_s \tag{5.29}$$

The actual evolution of the PSD during battery usage can be very complex and depends on many system properties (e.g. tendency of solvent to co-intercalate, surface properties, surface films, active material stiffness etc.). Due to this complexity, detailed experiments or simulations are needed to provide a quantitative analysis of electrode degradation. Since this is beyond the scope of this article, general scenarios and their impact on electrode performance are simulated.

5.3.2 Degradation scenarios

The impact of a hypothetical change of the PSD through agglomeration and cracking is analyzed. This degradation analysis does not aim to provide actual quantitative data of electrode degradation, but rather to point out the fundamental differences between both processes. Both scenarios are applied with constant kernals for particle cracking $\alpha^{cr} = 0.5$ in cracks simulation run time and agglomeration $\beta^{agl} = 0.5$ in $(n_s)^{-1}$ in agglomerations m^3 per degradation simulation run time. Furthermore, in the case of particle cracking, a disconnection of one part of the particle can be expected [144, 8]. For particle cracking, in this scenario the particle breaks into two same sized particle, while one particle is still electrically connected and the other one is disconnected, which leads to:

$$b^{\text{act}}(R_s|P_s) = b^{\text{dis}}(R_s|P_s) \tag{5.30}$$

$$b^{\text{act}}(R_s|P_s) = \begin{cases} 1 \text{ for } \frac{P_s}{2^{\frac{1}{3}}} = R_s \\ 0 \text{ for } \frac{P_s}{2^{\frac{1}{3}}} \neq R_s \end{cases} \tag{5.31}$$

The impact of the SEI can be neglected, because it is mainly formed out of the electrolyte and lithium provided by the positive electrode. This denotes that the SEI does impact the counter amount of cyclable lithium in the counter electrode, rather than the actual storage capacity of the graphite electrode, which is investigated here.

5.3.3 Discussion of electrode degradation

The previous results show that PSDs should have a considerable impact on electrode performance and degradation processes like solvent co-intercalation. This can cause graphite exfoliation or particle cracking [144, 8]. A restructuring of the electrode morphology, i.e. a change of the PSD, can impact electrode performance and result in decreasing battery capacity during usage. Figure 5.7 shows the resulting change of the PSD through introduced cracking and agglomeration scenarios, where the original PSD is referred to a reference PSD. If particle cracking occurs, the number of large particles decreases while the number of smaller particles increases. As theoretically and experimentally observed, this process could continue until the electrode particles are broken into dust [8]. In general, this leads to a PSD with a higher surface area to volume ratio, but also leads to loss of active material through electrical disconnection. Furthermore, a loss of cyclable lithium in the disconnected particles would cause a unbalancing of the battery system. The effect of an unbalanced system is not considered in this single electrode analysis, as presented here. Particle agglomeration causes a decrease of the number of smaller particles and increase of larger particles, but does not affect the active material fraction. Since active material content is constant, a change in electrode capacity is only caused by change of internal resistances, such as diffusion or surface reactions.

In Table 5.3, electrode capacity of a case before and after both degradation scenarios is shown for different initial particle shapes and constant particle scale of $\lambda = 0.2$. Electrode capacity is further presented dependent on C-rate to show the impact on electrode kinetic. It can be seen that for very low C-rates, only in case of particle cracking the electrode capacity decreases, because active material fraction for agglomeration stays constant and internal resistances are very low for low rates. In contrast, for higher rates both scenarios considerably impact electrode capacity. Discharge capacities after degradation are comparably lowered for high rates even so only for the case of particle cracking active material is lost. The effect of disconnected active material in the scenario of particle cracking is partly compensated through better electrode performance with smaller particles, due to shorter diffusion pathways and higher surface to volume ratio.

The results demonstrate that a change of PSD can be modeled using population balances as presented in this work. A realistic cracking and agglomeration needs to be setup and parameterized carefully to produce quantitative precise degradation scenarios. Nevertheless, in general, both degradation scenarios can lead to a decrease of electrode capacity, but due to completely different reasons, which can be quantified with the charge rate dependent capacity.

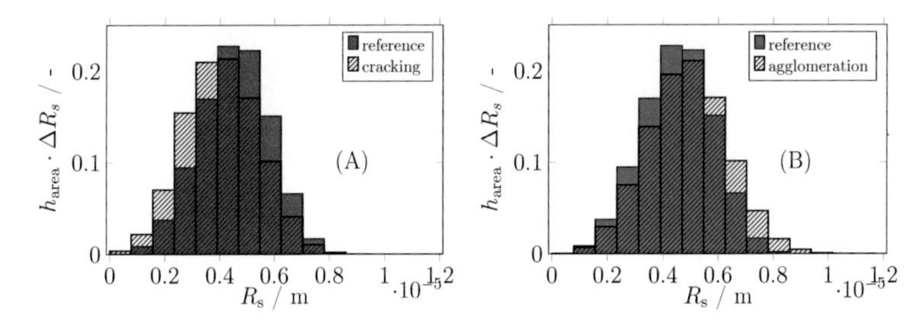

Figure 5.7: Number fraction density for a reference PSD and PSD after the change scenarios for particle cracking (A) and agglomeration (B). Reprinted from publication [60].

5.4 Concluding remarks

In this chapter, particle size was introduced as distributed parameter into the single particle electrode model for graphite electrodes to study the effect of PSD on perfomance and degradation.

The presented results show the impact of the shape and scale of the PSD on electrode performance. In general, small scaled PSDs with narrow distribution show lower internal resistances and smaller local surface current densities. This implies that they provide higher electrode capacity and lower probability for degradation processes related to local current densities such as solvent co-intercalation. It was further shown that in smaller scaled PSDs, diffusion processes determine the electrode resistance, while in larger distributions surface reactions are limiting through small specific surface areas. In general, the results indicate that for a graphite electrode, surface area or volume based mean approximation with commonly used single particle approach are accurate over a wide range of PSDs and only for coarse distributions they considerably differ from accurate considerations with multiple particle approaches. Therefore it can be concluded that as long as PSD is not exceedingly coarse, the surface area and volume based approximations are sufficiently accurate to predict electrode capacity for large and small scaled PSDs.

Analysis of the theoretical local current densities indicates that the highest current densities occur at large particles in coarse and large scaled distributions at the beginning of discharge. Since large current densities can lead to degradation through restructuring of electrode morphology, a population balance based approach was provided to investigate the effect of a change of the PSD. Exemplary scenario based simulations show that both particle agglomeration and particle cracking can lead to a decrease of electrode capacity, while the fundamental reasons for the loss of active material and increase of internal resistance can be very different.

Table 5.3: Electrode discharge capacity with different C-rates for reference scenarios with different initial PSD shape and degradation scenarios for cracking and agglomeration. Reprinted from publication [60].

shape	scenario	0.01 C	0.1 C	1 C
	reference	**0.995**	**0.951**	**0.58**
$k = 8$	agglomeration.	0.993	0.937	0.518
	cracking	0.774	0.734	0.419
	reference	**0.994**	**0.94**	**0.537**
$k = 4$	agglomeration.	0.992	0.927	0.484
	cracking	0.773	0.724	0.387
	reference	**0.988**	**0.882**	**0.385**
$k = 2$	agglomeration.	0.986	0.869	0.35
	cracking	0.768	0.673	0.274
	reference	**0.978**	**0.802**	**0.272**
$k = 1.5$	agglomeration.	0.975	0.787	0.25
	cracking	0.759	0.604	0.192

Based on these results, it is suggested to consider the PSD effect when analysing degradation processes in lithium-ion batteries and its effect on performance. Both the uneven degradation at differently sized particles due to the diversity of the surface current density and the change of the distribution during battery operation, are thereby of particular interest. A logical next step would be to combine a model as presented in chapters 3 and 4 with accurate consideration of uneven surface film growth as a consequence of particle size distribution heterogeneity. Further, future developments could account for the degradation of the electrode structure and its impact on film reformation at exposed surface areas during particle cracking or disconnection.

Chapter 6

Conclusions

6.1 Summery

The objective of this thesis was the development of methods for multiscale simulation of degradation in lithium-ion batteries. In chapter 1, Section 1.2, the multiscale nature of battery degradation was outlined. It was indicated that there are interactions between main processes, side reactions, film growth and electrode restructuring throughout the scales, i.e. electrode scale, particle scale, mesoscale, and atomistic scale. However, those aspects are commonly modeled independently and on single scales. In this thesis, a novel multiscale methodology has been developed, which bridges a wide range length scales, and enables to study interactions as outlined in chapter 1. The thesis provides algorithms, models, and concepts and applies them for analysis of multiscale effects in lithium-ion batteries.

In chapter 2, concept and algorithms for direct coupling of continuum and kMC models, i.e. multiparadigm simulation, were developed and analyzed with respect to accuracy and computational cost. These methods were used in chapter 3 to develop a multiparadigm modeling approach for simulation of heterogeneous surface film growth problems in lithium-ion batteries. In chapter 4, the model concept of chapter 3 was extended to reveal the evidence of multiscale effects in film formation in an actual technical cell. As models in chapter 3 and 4 neglect electrode heterogeneity and do not cover electrode restructuring, in chapter 5 heterogeneous models and their degradation have been investigated by introducing particle size distribution and population balances into commonly applied homogeneous models. With this, chapters 2-5 provide a comprehensive methodology to cover most of the interactions outlined in Sections 1.2. The major achievements of this thesis are illustrated in Figure 6.1. The most relevant findings will be summarized in the following.

Degradation often affects processes on several scales. For instance, growth of surface films through unwanted side reactions at electrochemically active surfaces reduces porosity on electrode scale, increases film resistance on mesoscale scale, and changes

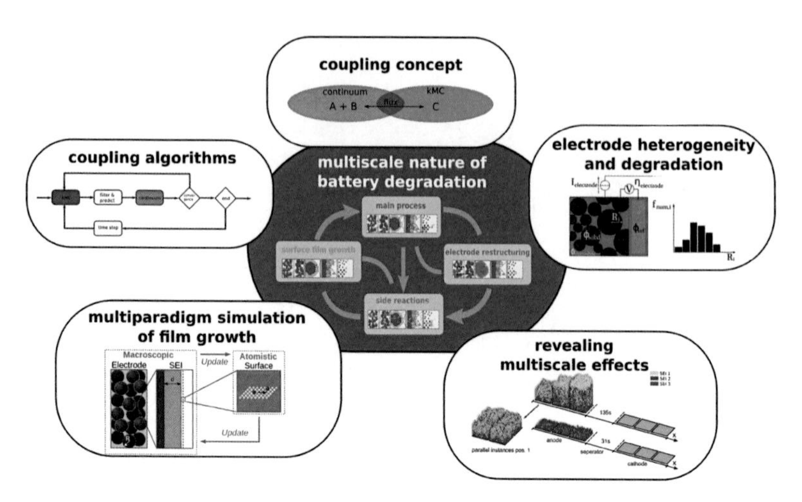

Figure 6.1: Major achievements of this thesis.

the structure of adsorption sites through deposition of solid components on atomistic scale. In chapter 1, it has been shown that currently available models do either cover processes on electrode to particle scale or meso- to atomistic scale, while different modeling paradigms need to be applied. This demonstrated that there is a need for coupling different modeling paradigms in order to analyze multiscale effects. In this thesis it has been shown that direct coupling of continuum and kMC models is a very powerful approach to bridge these wide range of length scales.

The coupling of continuum and kMC models is challenging. In continuum models time is continuous, i.e. time dependency is described by deterministic differential equations. In kMC models however, time is discrete, i.e. system states are only defined at discrete points in time, and the mathematical description is not deterministic, i.e. based on random numbers. Furthermore, the computational cost is significantly higher, i.e. limited to shorter time scales. Direct coupling of both simulations requires synchronization of both model types, which includes handling of stochastic fluctuations of the kMC and assuring convergence between the models. In this thesis, a concept and adequate algorithms were developed and tested. Here, kMC simulations are only applied for heterogeneous process on electrochemical active surfaces, while all other processes and species are covered by the continuum model. Every species is covered by one of the modeling methods and enters the other simulation code via a process flux. The flux thereby is synchronized using the developed MPAs. With this, all issues outlined above were solved.

Using the developed MP methodology, common continuum battery models, e.g. SP-models and P2D models, were extended to include heterogeneous surface film growth mechanisms. This was realized by describing surface film growth with a solid-on-solid kMC model. With this model, hypothetical reaction mechanisms at the surface film/ electrolyte can be included into a full cell or electrode simulation. The simulation results could reproduce experimental observation from literature and electrochemical experiments. This includes: the slope of potential during the first charge, reported film compositions in EC based electrolytes, and impact of particle size on reversible capacity. This shows for the first time a model which is capable of simulating complex side reaction and film growth mechanism from the electrode to the atomistic scale.

The modeling approach has been used to investigate multiscale effects within an actual technical cell. A step-wise parameter identification strategy facilitates the separation of main and side reaction processes. Analysis of coupled simulations revealed the evidence of multiscale interaction, which were shown to be particularly relevant with very slow and very fast formation procedures. Here, it could be seen that the structure of the surface film considerably depends on the charging rate. Moreover, due to spacial distribution of concentration and electrical potentials, the structure can significantly vary within an electrode. The presented results show that such multiscale effects need to be considered to identify optimal charging strategies to achieve desired film structures. As the presented model is the only presently reported model which directly couples heterogeneous film growth with continuum cell models, this interdependence could be indicated and analyzed for the first time.

Common battery models assume homogeneity on the electrode scale. In this thesis, heterogeneity on the electrode scale has been included by using statistically distributed parameters, i.e. particle size distribution. Simulation results showed, that in the case of coarse particle size distribution, i.e. significant heterogeneity, applied mean approximations do not usually allow the reproduction of electrochemical behavior, i.e. potential and capacity during discharge. Further, it can be seen that local conditions, i.e. concentration at particle boundaries and surface current density, depend on the particle size, while the highest surface current densities were observed at large particles at the beginning of discharge. Results indicate that degradation, e.g particle cracking or surface film growth, may not only depend on the position within the electrode, but also on the local particle size.

Usual battery ageing models treat degradation as a homogeneous process. In those models, mechanical degradation or surface film growth can depend on spacial position within the electrode, but not on local heterogeneity. In this thesis, for the first time, population balances have been applied to consider heterogeneous degradation. By presenting simulation scenarios for electrode cracking and agglomeration, the application of this approach for simulation of electrode restructuring has been demonstrated and verified.

It has been shown that the modeling methods developed in this thesis cover most of the interactions outlined in chapter 1 and enables bridging of the scales from electrode to atomistic in direct coupled simulations.

6.2 Future challenges

Though this thesis provides a comprehensive methodology for multiscale modeling of degradation in lithium-ion batteries, several challenges remain which should be addressed in future research. There is a need to refine parameter identification and multiscale analysis strategies in order to systematically optimize electrolytes and formation procedures. Moreover, models should be extended and further interconnected to also cover other important interactions and multiscale effects. Finally, there is a need to connect this approach to other simulation methods on even smaller scales and integrate them into an easy-to-use simulation environment.

The identification of model parameters, as given in chapter 4, provided basic steps based on a quite limited set of experimental data. The results prove that the multiscale model is adequate to quantitavely describe electrochemical measurements, which proves the general validity of the model. However, based on this data set, parameters of almost every reaction mechanism could be adapted to fit experimental data. This denotes that no reliable identification of a mechanism and its parameters is presently possible. To resolve this issue, the first step should be to reduce the complexity of investigated problem by using a plain, one solvent, electrolyte. Another strategy is to increase applied experimental data. As shown in chapter 4, for electrochenmical characterization the focus should be laid on first charge discharge cycles, as here film growth processes are actually detectable. The potential plateau observed in the experiment will shift depending on the applied current due to activation energy of the rate limiting processes. This could enable the identification of activation energies. Nevertheless, for the example of methanol oxidation, it has been shown that several mechanisms and parameter sets can yield similar behavior at constant current [162]. The authors applied dynamic measurements to gain further information and identify reaction mechanisms. As during the first cycles, the system undergoes rapid and fundamental changes, measurement procedure should be fast and limited to particular frequencies of interest. To further validate reaction mechanisms, data about concentration of electrolyte components, gas species or solid components are needed. Gas species are detectable online by special experimental setups using a differential electrochemical mass spectrometer as shown for CO oxidation [163] and batteries [164]. Finally, solid and electrolyte component concentrations should be measured in a post mortem analysis at different stages of the formation processes. To conclude, in order to validate particular reaction mechanisms there is a need to systematically increase the experimental data and to first focus on plain electrolytes.

Typically, parameters are identified using mathematical optimization methods. However, simulation time for multiscale simulations are rather high, i.e. approximatly 12h for simulation of the formation process, thus the identification of a large set of parameters using such methods is highly challenging. Therefore, there is a need to signficantly reduce the computational cost of coupled simulations. This could be faciltitated, for instance, by shifting some processes back to continuum simulation during parameter identification procedures or by application of coarse graining methods [93].

In chapter 5, it has been shown how the heterogeneity of the electrode structure can be considered. Further, it was shown that this heterogeneity can have considerable impact on local conditions and thus on the surface film growth. Therefore, heterogeneous models need to be coupled with kMC simulations of surface film growth, using the presented MP approaches. Here, film growth in the heterogeneous electrodes can be treated by population balance equations, as shown in chapter 5. Moreover, interaction between surface film growth and electrode restructuring is strongly related to film breakage and reformation cycles, which has been discussed in chapter 1. Particle cracking or particle disconnection can lead to exposed or uncovered electrochemically active surface area. On these spots, rapid film growth will be triggered. High local current rates may lead to further damage of electrode structure. The multiscale interaction described is currently not covered by adequate models. However, understanding this aspect is crucial to ensure safety and long life time of batteries.

In chapter 3 it has been demonstrated that atomistic mechanisms and parameters, e.g. activation energies, can be introduced into the kMC model. By systematically transferring knowledge from those methods, systems and materials, e.g. electrolyte additives, could be studied from a first principle. This would significantly reduce cost and time for product development, as screening of electrolyte additives presently requires a huge amount of expensive experiments. However, present results from DFT or MD simulations can barely be used for evaluations on electrode scale. In general, with the methodology for MP simulation given in this thesis, an important first step towards first principle multiscale simulation has been completed. Future research should build upon this and address the development of a software package which integrates DFT, MD, kMC and continuum models in seamless simulation environment for multiscale simulation of batteries.

In this thesis, the developed method has been applied exclusively to degradation in lithium-ion batteries, and thereby has focused on issues connected to the SEI.

However, the method concept could be transferred to related degradation processes in lithium-ion batteries, all solid state batteries, fuel cells, or other electrochemical systems.

6.3 Impact

To sum up, with this work, significant physical insight into degradation and multiscale interaction was given. Moreover, the work provides a large set of novel algorithms, models and methods for multiscale analysis of lithium-ion batteries. This opens new opportunities of research and development in this and related fields.

In this work, processes in lithium-ion batteries from the electrode to the atomstic scale have been connected for the first time. This will significantly contribute to the vision to develop batteries from a first principle. Presented results provide a deeper understanding of degradation, i.e. film growth and electrode restructuring and in particular interactions through the scales. Application of this comprehensive approach will enable the optimization of batteries, their materials, and production procedures. This will increase life time, ensure safety, and reduce cost of lithium-ion batteries, and thus significantly contributes to transfer the energy market towards renewable energies and in particular to the development towards electromobility.

Appendix A

Parameter Tables

All parameters applied for simulations are summarized in this chapter. In the caption of the tables the chapter is indicated accordingly. It should be noted that even values of alleged equal parameter may deviate in between the chapters. Some alleged physical parameters can deviate in between different cells as they are indeed effective parameters, which need to be adapted to describe particular systems. Further, if parameters have been identified at electrochemical experiments, they were not tested for uniqueness, since this is out of the scope of this thesis.

Table A.1: Model parameters (applied in chapter 2). Reprinted from publication [89] with permission from Elsevier.

Parameter	Value
ideal gas constant R [J mol^{-1} K^{-1}]	8.314
temperature T [K]	300
Faraday constant [A s mol^{-1}]	96485.33289
specific surface area a_s [m^{-1}]	1×10^6
double layer capacitance C^{DL} [F m^{-2}]	0.2
distance between lattice sites ΔL [m]	6×10^{-10}
site density N_s [m^{-2}]	$1/\Delta L^2$
site-occupancy number o_s [mol^{-1}]	6.022×10^{23}
input amplitude $\bar{A}$ [A m^{-2}]	1
input frequency f [s^{-1}]	0.1
symmetry factor β [–]	0.5
rate constant k_{I} [s^{-1}]	1×10^{13}
rate constant k_{II} [s^{-1}]	1×10^{10}
rate constant k_{III} [s^{-1}]	1×10^{10}
rate constant k_{IV} [s^{-1}]	1×10^{13}
activation energy E_{I}^{A} [kJ mol^{-1}]	45
activation energy E_{II}^{A} [kJ mol^{-1}]	50
activation energy E_{III}^{A} [kJ mol^{-1}]	50
activation energy E_{IV}^{A} [kJ mol^{-1}]	45
standard state Gibbs free energy $\Delta G_{\mathrm{I}}^{0}$ [kJ mol^{-1}]	0
standard state Gibbs free energy $\Delta G_{\mathrm{II}}^{0}$ [kJ mol^{-1}]	-0.965
standard state Gibbs free energy $\Delta G_{\mathrm{III}}^{0}$ [kJ mol^{-1}]	-0.965
standard state Gibbs free energy $\Delta G_{\mathrm{IV}}^{0}$ [kJ mol^{-1}]	1.93
standard state concentration of species A^{+}(E) $C_{\mathrm{A}+(E)}^{0}$ [mol m^{-3}]	1000
standard state concentration of species C(E) $C_{\mathrm{C}(E)}^{0}$ [mol m^{-3}]	1000
initial potential $\Delta\Phi(t=0)$ [V]	0
initial concentration of species $c_{\mathrm{A+(E)}}(t=0)$ [mol m^{-3}]	1000
initial concentration of species $c_{\mathrm{C(E)}}(t=0)$ [mol m^{-3}]	1000
initial surface fraction of species $\theta_{\mathrm{A+(ads)}}(t=0)$ [–]	0.18
initial surface fraction of species $\theta_{\mathrm{B(ads)}}(t=0)$ [–]	0.26
initial surface fraction of species $\theta_{\mathrm{C(ads)}}(t=0)$ [–]	0.38

Table A.2: Reaction processes including reaction rate constant k, Gibbs free energy ΔG^0, and activation energy E^A (applied in chapter 3). Reprinted from publication [106].

Number	Reactions	A [s^{-1}]	E^A [kJ mol^{-1}]	β	method
1	$C_3H_4O_3 + e^- \rightleftharpoons C_3H_4O_3^-$	5×10^{12}	65.27 [80]	0.5	kMC
2	$C_3H_4O_3 + e^- + Li^+ \rightleftharpoons LiC_3H_4O_3$	5×10^{12}	42.68 [80]	0.5	kMC
3	$C_3H_4O_3{}^- + e^- \rightleftharpoons CO_3^{2-} + C_2H_4$	5×10^{12}	275.31 [80]	0.5	kMC
4	$LiC_3H_4O_3 + e^- \rightleftharpoons LiCO_3^- + C_2H_4$	5×10^{12}	53 (chosen)	0.5	kMC
5	$C_3H_4O_3^- + Li^+ \rightleftharpoons LiC_3H_4O_3$	1×10^{13}	40 (chosen)	–	kMC
6	$CO_3^{2-} + Li^+ \rightleftharpoons LiCO_3^-$	1×10^{13}	40 (chosen)	–	kMC
7	$2\ LiC_3H_4O_3 \rightarrow (CH_2OCO_2Li)_2 + C_2H_4$	1×10^{13}	5 (chosen)	–	kMC
8	$LiCO_3^- + Li^+ + C_3H_4O_3 \rightarrow (CH_2OCO_2Li)_2$	1×10^{13}	80 (chosen)	–	kMC
9	$LiCO_3^- + Li^+ \rightarrow Li_2CO_3$	1×10^{13}	70 (chosen)	–	kMC
10	$Li(s) \rightleftharpoons V(s) + Li^+(s) + e^-(s)$	1×10^{13}	30	0.5	continuum
11	$Li^+(s) \rightleftharpoons Li^+(ads)$	1×10^{13}	30	–	continuum
12	$Li^+(ads) \rightleftharpoons Li^+(e)$	1×10^{13}	30	0.5	continuum
13	$PF_6^-(ads) \rightleftharpoons PF_6^-(e)$	1×10^{13}	30	0.5	continuum
14	$C_3H_4O_3(ads) \rightleftharpoons C_3H_4O_3(e)$	1×10^{13}	30	–	continuum
15	$2\ C_3H_4O_3(ads) + 2\ Li^+(ads) + 2\ e^-(ads) \rightarrow (CH_2OCO_2Li)_2 + C_2H_4(e)$	kMC input	–	–	continuum
16	$C_3H_4O_3(ads) + 2\ Li^+(ads) + 2\ e^-(ads) \rightarrow Li_2CO_3 + C_2H_4(e)$	kMC input	–	–	continuum
17	$C_3H_4O_3(ads) + Li^+(ads) + e^-(ads) \rightarrow LiC_3H_4O_3(e)$	kMC input	–	–	continuum

Table A.3: Standard chemical potential μ^0, surface diffusion coefficient D, binding energy J_i to solid i in kJ/mol, and desorption rate constant k^{des} as used in the kMC model (applied in chapter 3). Reprinted from publication [106].

Species	μ^0 [kJ mol^{-1}]	E^A_{bond}(LC); E^A_{bond}(LEDC); E^A_{bond}(a) [kJ mol^{-1}]	k^{des} [s^{-1}]
$C_3H_4O_3$	0	–	–
PF_6^-	0	–	–
Li^+	10	–	–
e^-	0	–	–
$C_3H_4O_3^-$	−33.93	22; 22; 22	–
CO_3^{2-}	256.44	22; 22; 22	–
$LiC_3H_4O_3$	−574.63	8; 27; 29	5×10^7
$LiCO_3^-$	−780.48	27; 3; 29	–
C_2H_4	0	5; 5; 5	5×10^7
$(CH_2OCO_2Li)_2$	−1386.91	–	–
Li_2CO_3	−1399	–	–

Table A.4: Standard chemical potential μ^0 in kJ/mol and standard concentration C^0 in mol/m^3 (applied in chapter 3). Reprinted from publication [106].

Species	μ^0 [kJ mol^{-1}]	C^0 [mol m^{-3}]
Li(s)	−11.65 [117]	c_{max}
e^-(s)	0	1
V(s)	0	c_{max}
Li^+(film)	–	1000
Li^+(ads)	10	N_s/o_s
PF_6^-(ads)	0	N_s/o_s
V(ads)	−1	N_s/o_s
$C_3H_4O_3$(ads)	0	N_s/o_s
Li^+(ads)	0	1200
PF_6^-(e)	0	1200
$C_3H_4O_3$(e)	0	15000

Table A.5: Other model parameters (applied in chapter 3). Reprinted from publication [106].

Parameter	value
Maximum concentration in the solid, c_{max} [mol m^{-3}]	16100
temperature, T [K]	300
solid diffusion coefficient, $D_s^{\mathrm{Li(s)}}$ [m^2 s^{-1}]	1×10^{-14}
electrical resistivity of SEI, R^{film} [Ω m]	5×10^5
double layer capacitance at s/film interface, $C_{\mathrm{s,film}}^{\mathrm{DL}}$ [F m^{-2}]	0.2
double layer capacitance at ads/e interface, $C_{\mathrm{ads,e}}^{\mathrm{DL}}$ [F m^{-2}]	10
active material volume fraction, ε_s [–]	0.58
concentration of lithium ions in electrolyte, $c_e^{\mathrm{Li}^+(e)}$ [mol m^{-3}]	1200
start concentration of lithium in solid $c_s^{\mathrm{Li}(s)}(0)$ [mol m^{-3}]	0
active material particle radius, R_s [m]	3×10^{-6}, 10×10^{-6}
site-occupancy number, o_s [mol^{-1}]	6.022×10^{23}
site density N_s [m^{-2}]	$1/(\Delta L^2)$
C-rate (based on concentration at 0 V) [–]	0.1
surface area roughness factor, r_s [–]	5
kMC lattice size, ΔL [m]	6×10^{-10}
cutoff voltage, E_{cut} [V]	0
specific electron leakage activation energy, $\hat{E}_{\mathrm{film}}^{A}$ [J m^{-1}]	1×10^{12}
surface diffusion coefficient, D [m^2 s^{-1}]	1×10^{-13}

Table A.6: Chemical species and their standard state chemical potentials (applied in chapter 4). Reproduced from publication [1] with kind permission from Wiley-VCH Verlag GmbH & Co. KGaA.

Species	μ^0 [kJ] (F^{-1}[V])			method	source
	anode	separator	cathode		
S_1(ads)	0 (0)	–	0 (0)	continuum	chosen
S_1(e)		0 (0)		continuum	chosen
S_2(ads)	0 (0)	–	0 (0)	continuum	chosen
S_2(e)		0		continuum	chosen
S_3(ads)	0 (0)	–	0 (0)	continuum	chosen
S_3(e)		0 (0)		continuum	chosen
Li(s)	−12.42 (−0.1287)	–	−375.89 (−3.8958)	continuum	fit
Li^+(ads)	0 (0)	–	−96.49 (−1)	continuum	chosen
Li^+(e)		0		continuum	chosen
V(s)	0 (0)	–	0 (0)	continuum	fit
V(ads)	−3.86 (−0.04)	–	−3.86 (−0.04)	continuum	chosen
LiS_1(ads)	−56.93 (−0.59)	–	−56.93 (−0.59)	kMC	chosen
LiS_1(e)		−55 (−0.57)		continuum	chosen
SEI_1(film)	−154.38 (−1.6)	–	154.38 (−1.6)	kMC	fit
SEI_2(film)	−77.19 (−0.8)	–	−77.19 (−0.8)	kMC	fit
SEI_3(film)	−57.89 (−0.6)	–	−57.89 (−0.6)	kMC	fit
SEI_1(e)		−154.38 (−1.6)		continuum	fit

Table A.7: Redlich Kister coefficients for anode and cathode (applied in chapter 4). Reproduced from publication [1] with kind permission from Wiley-VCH Verlag GmbH & Co. KGaA.

Coefficient	Anode	Cathode	source
A_1^{RK}	−3797	−29626	fit
A_2^{RK}	5260	13328	fit
A_3^{RK}	−7290	−7942	fit
A_4^{RK}	8174	−4926	fit
A_5^{RK}	−3165	6859	fit
A_6^{RK}	−45	0	fit
A_7^{RK}	2079	0	fit
A_8^{RK}	−2388	0	fit
A_9^{RK}	1644	0	fit
A_{10}^{RK}	27	0	fit
A_{11}^{RK}	−2230	0	fit
A_{12}^{RK}	4818	0	fit
A_{13}^{RK}	−7585	0	fit
A_{14}^{RK}	10439	0	fit
A_{15}^{RK}	−13277	0	fit
A_{16}^{RK}	16050	0	fit
A_{17}^{RK}	−18709	0	fit

Table A.8: Battery parameters Part I (applied in chapter 4). Reproduced from publication [1] with kind permission from Wiley-VCH Verlag GmbH & Co. KGaA.

Parameter	Anode	Separator	Cathode	Source
double layer capacitance $C^{DL}_{\mathrm{s,film}}$ [F m^{-2}]	0.1	–	0.2	chosen
double layer capacitance $C^{DL}_{\mathrm{ads,e}}$ [F m^{-2}]	0.1	–	0.2	chosen
maximal concentration in solid $c_{\max}$ [mol m^{-3}]	32000	–	30000	chosen
roughness factor r_s [–]	10	–	1	chosen
thickness δ [m]	55.25×10^{-6}	20×10^{-6}	60×10^{-6}	measured
solid diffusion coefficient D_s^{Li} [m^2 s^{-1}]	1.17×10^{-13}	–	7.3294×10^{-16}	fit
diffusion coefficient of S_1 $D_e^{S_1}$ [m^2 s^{-1}]		1×10^{-13}		chosen
diffusion coefficient of S_2 $D_e^{S_2}$ [m^2 s^{-1}]		1×10^{-12}		chosen
diffusion coefficient of S_3 $D_e^{S_3}$ [m^2 s^{-1}]		1×10^{-14}		chosen
diffusion coefficient of LiS_1 $D_e^{LiS_1}$ [m^2 s^{-1}]		0		chosen
coefficients for salt diffusion (b_1,b_2,b_3) [mol,m,s]	(2.5×10^{-10},	-1.63×10^{-13},	3.183×10^{-17})	fit
surface diffusion of LiS_1 $D_{\mathrm{ads}}^{LiS_1}$ [m^2 s^{-1}]	1×10^{-14}	–	1×10^{-14}	chosen
transference number t_p [–]		0.3		chosen
tortuosity τ_e [–]	13.8918	3.03	2.3041	fit
solid volume fraction ε_s [–]	0.65	0.5	0.6	measured
solid electrical conductivity σ_s [S m^{-1}]	672.416	–	5.4179	fit
electrical resistance surface film R^{film} [Ω m]		4×10^{6}		chosen
particle radius R_s [m]	5.5×10^{-6}	–	11.5×10^{-6}	material info.
temperature T [K]		300		chosen
site density N_s [m^{-2}]		2.77778×10^{18}		chosen
site occupancy number o_s [mol^{-1}]		6.022×10^{23}		chosen
grid size ΔL [m]		6×10^{-10}		chosen
standard state concentration Li^+(e) $C^0_{Li^+(e)}$ [mol m^{-3}]		1000		chosen
standard state concentration LiS_1(e) $C^0_{LiS_1(e)}$ [mol m^{-3}]		1000		chosen
standard state concentration SEI_1(e) $C^0_{SEI_1(e)}$ [mol m^{-3}]		1000		chosen
standard state concentration S_x $C^0_{S_x(e)}$ [mol m^{-3}]		4000		chosen

Table A.9: Battery parameters Part II (applied in chapter 4). Reproduced from publication [1] with kind permission from Wiley-VCH Verlag GmbH & Co. KGaA.

Parameter	Anode	Separator	Cathode	Source
pre exponential factor A [s^{-1}]	1×10^{10}	–	1×10^{10}	chosen
activation energy E^A_{I} [kJ] (F^{-1}[V])	2.89 (0.03)	–	−132474.36 (−1.373)	fit
activation energy E^A_{II} [kJ] (F^{-1}[V])	18.33 (0.19)	–	18.33 (0.19)	chosen
activation energy E^A_{III} [kJ] (F^{-1}[V])	19.3 (0.20)	–	19.3 (0.20)	chosen
activation energy E^A_{IV} [kJ] (F^{-1}[V])	20.26 (0.21)	–	20.26 (0.21)	chosen
activation energy E^A_{V} [kJ] (F^{-1}[V])	1.93 (0.02)	–	1.93 (0.02)	chosen
activation energy E^A_{VI} [kJ] (F^{-1}[V])	0.96 (0.01)	0.96 (0.01)	0.96 (0.01)	chosen
activation energy E^A_{VII} [kJ] (F^{-1}[V])	33.77 (0.35)	–	33.77 (0.35)	chosen
activation energy E^A_{VIII} [kJ] (F^{-1}[V])	28.95 (0.3)	–	28.95 (0.3)	chosen
activation energy E^A_{XI} [kJ] (F^{-1}[V])	28.95 (0.3)	–	28.95 (0.3)	chosen
activation energy E^A_{X} [kJ] (F^{-1}[V])	28.95 (0.3)	–	28.95 (0.3)	chosen
activation energy E^A_{XI} [kJ] (F^{-1}[V])	28.95 (0.3)	–	28.95 (0.3)	chosen
specific activation energy $\hat{E}^A_{\mathrm{film}}$ [kJ m^{-1}] (F^{-1}[V m^{-1}])	8.29×10^8 (8.6×10^6)	–	8.29×10^8 (8.6×10^6)	fit
bonding energy $E^A_{\mathrm{bond}}(\mathrm{LiS_1}\vert\mathrm{SEI_1})$ [kJ] (F^{-1}[V])	7 (0.0725)	–	7 (0.0725)	chosen
bonding energy $E^A_{\mathrm{bond}}(\mathrm{LiS_1}\vert\mathrm{SEI_2})$ [kJ] (F^{-1}[V])	3 (0.0311)	–	3 (0.0311)	chosen
bonding energy $E^A_{\mathrm{bond}}(\mathrm{LiS_1}\vert\mathrm{SEI_3})$ [kJ] (F^{-1}[V])	3 (0.0311)	–	3 (0.0311)	chosen
bonding energy $E^A_{\mathrm{bond}}(\mathrm{LiS_1}\vert\mathrm{solid})$ [kJ] (F^{-1}[V])	7 (0.0725)	–	7 (0.0725)	chosen
bonding energy $E^A_{\mathrm{bond}}(\mathrm{SEI_1}\vert\mathrm{SEI_1})$ [kJ] (F^{-1}[V])	12 (0.1244)	–	12 (0.1244)	chosen
bonding energy $E^A_{\mathrm{bond}}(\mathrm{SEI_2}\vert\mathrm{SEI_3})$ [kJ] (F^{-1}[V])	−4 (−0.0415)	–	−4 (−0.0415)	chosen

Table A.10: Model parameters (applied in chapter 5). Reprinted from publication [60].

Parameter	value
temperature T / K	300 (chosen)
solid diffusion coefficient $D_s^{\mathrm{Li(s)}}$ / m^2 s^{-1}	$1 \cdot 10^{-15}$ (based on section 4.1)
ideal exchange current density rate k_{ct} / m$^{2.5}$ mol$^{-0.5}$ s^{-1}	$1.429 \cdot 10^{-9}$ [117]
symmetry factor β / -	0.5 (chosen)
double layer capacitance C^{DL} / F m^2	0.2 [134]
active material volume fraction ε_s / -	0.6 (chosen)
standard state chemical potential of lithium in solid μ^0_{Lis} / J mol^{-1}	$-1.165 \cdot 10^4$[117]
standard state chemical potential of lithium-ion in electrolyte $\mu^0_{\mathrm{Li^+(e)}}$ / J mol^{-1}	0[117]
standard state chemical potential of vacancy in solid $\mu^0_{\mathrm{V(s)}}$ / J mol^{-1}	0[117]
standard state chemical potential of electron $\mu_{\mathrm{e^-(s)}}$ / J mol^{-1}	0[117]
concentration lithium-Ions in electrolyte $c_{\mathrm{Li^+(e)}}$ / mol m^{-3}	1200 [134]
maximum concentration of lithium and vacancies in solid c_{max} / mol m^{-3}	16100 [117]
start concentration of ithium in solid $c_s^{\mathrm{Li(s)}}(0)$ / mol m^{-3}	13098 (chosen)
maximum theoretical discharge capacity C_{Ah} / Ah	$c_s^{\mathrm{Li(s)}}(0)\mathrm{F}\varepsilon_s$
scale parameters λ / µm^{-1}	0.8, 0.4, 0.2, 0.1, 0.05 (chosen)
shape parameters k / -	8, 4, 2, 1.5 (chosen)

Bibliography

[1] F. Röder, V. Laue, and U. Krewer. Model Based Multiscale Analysis of Film Formation in Lithium-Ion Batteries. *Batteries and Supercaps*, 2:248–265, 2019.

[2] L. Bodenes, R. Naturel, H. Martinez, R. Dedryvère, M. Menetrier, L. Croguennec, J.-P. Pérès, C. Tessier, and F. Fischer. Lithium Secondary Batteries Working at Very High Temperature: Capacity Fade and Understanding of Aging Mechanisms. *Journal of Power Sources*, 236:265–275, 2013.

[3] M. Reichert, D. Andre, A. Rösmann, P. Janssen, H. G. Bremes, D. U. Sauer, S. Passerini, and M. Winter. Influence of Relaxation Time on the Lifetime of Commercial Lithium-Ion Cells. *Journal of Power Sources*, 239:45–53, 2013.

[4] A. Barré, B. Deguilhem, S. Grolleau, M. Gérard, F. Suard, and D. Riu. A Review on Lithium-Ion Battery Ageing Mechanisms and Estimations for Automotive Applications. *Journal of Power Sources*, 241:680–689, 2013.

[5] M. Broussely, Ph Biensan, F. Bonhomme, Ph Blanchard, S. Herreyre, K. Nechev, and R. J. Staniewicz. Main Aging Mechanisms in Li Ion Batteries. *Journal of Power Sources*, 146(1-2):90–96, 2005.

[6] M. Dubarry and B. Y. Liaw. Identify Capacity Fading Mechanism in a Commercial $LiFePO_4$ Cell. *Journal of Power Sources*, 194(1):541–549, 2009.

[7] J. Vetter, P. Novák, M. R. Wagner, C. Veit, K. C. Möller, J. O. Besenhard, M. Winter, M. Wohlfahrt-Mehrens, C. Vogler, and A. Hammouche. Ageing Mechanisms in Lithium-Ion Batteries. *Journal of Power Sources*, 147(1-2):269–281, 2005.

[8] D. Aurbach, B. Markovsky, I. Weissman, E. Levi, and Y. Ein-Eli. On the Correlation between Surface Chemistry and Performance of Graphite Negative Electrodes for Li Ion Batteries. *Electrochimica Acta*, 45(1):67–86, 1999.

[9] W.-R. Liu, Z.-Z. Guo, W.-S. Young, D.-T. Shieh, H.-C. Wu, M.-H. Yang, and N.-L. Wu. Effect of Electrode Structure on Performance of Si Anode in Li-Ion Batteries: Si Particle Size and Conductive Additive. *Journal of Power Sources*, 140(1):139–144, 2005.

[10] Z. Wu, W. Ren, L. Wen, L. Gao, J. Zhao, Z. Chen, G. Zhou, F. Li, and H.-M. Cheng. Graphene Anchored with Co_3O_4 Nanoparticles as Anode of Lithium Ion Capacity and Cyclic Performance. *ACS Nano*, 4(6):3187–3194, 2010.

[11] H. Li, X. Huang, L. Chen, G. Zhou, Z. Zhang, D. Yu, Y. J. Mo, and N. Pei. The Crystal Structural Evolution of Nano-Si Anode Caused by Lithium Insertion and Extraction at Room Temperature. *Solid State Ionics*, 135:181–191, 2000.

[12] U. Krewer, F. Röder, E. Harinath, R. D. Braatz, B. Bedürftig, and R. Findeisen. Dynamic Models of Li-ion Batteries for Diagnosis and Operation - A Review and Perspective. *Journal of the Electrochemical Society*, 165:A3656–A3673, 2018.

[13] A. A. Franco. Multiscale Modelling and Numerical Simulation of Rechargeable Lithium Ion Batteries: Concepts, Methods and Challenges. *RSC Adv.*, 3:13027–13058, 2013.

[14] R. D. Braatz, E. G. Seebauer, and R. C. Alkire. Multiscale Modeling and Design of Electrochemical Systems. In *Electrochemical Surface Modification: Thin Films, Functionalization and Characterization*, pages 289–334. Wiley-VCH, 2008.

[15] J. Newman and K. E. Thomas-Alyea. *Electrochemical Systems*. John Wiley & Sons, New York, 2012.

[16] J. Newman and W. Tiedemann. Porous-Electrode Theory with Battery Applications. *AIChE Journal*, 21(1):25–41, 1975.

[17] K. Reuter. *First-Principles Kinetic Monte Carlo Simulations for Heterogeneous Catalysis First-Principles Kinetic Monte Carlo Simulations for Heterogeneous Catalysis: Concepts , Status and Frontiers*, pages 71–111. Wiley-VCH, 2011.

[18] A. Chatterjee and D. G. Vlachos. An Overview of Spatial Microscopic and Accelerated Kinetic Monte Carlo Methods. *Journal of Computer-Aided Materials Design*, 14(2):253–308, 2007.

[19] W. G. Hoover. *Molecular Dynamics*, volume 258 of *Lecture Notes in Physics, Berlin Springer Verlag*. Springer, 1986.

[20] E. K. Gross and R. M. Dreizler. *Density Functional Theory*, volume 337. Springer Science & Business Media, 2013.

[21] M. Doyle, T. F. Fuller, and J. Newman. Modeling of Galvanostatic Charge and Discharge of the Lithium/Polymer/Insertion Cell. *Journal of The Electrochemical Society*, 140(6):1526–1533, 1993.

[22] T. F. Fuller, M. Doyle, and J. Newman. Simulation and Optimization of the Dual Lithium Ion Insertion Cell. *Journal of The Electrochemical Society,*, 141(1):1–10, 1994.

[23] E. Gongadze and A. Iglič. Asymmetric Size of Ions and Orientational Ordering of Water Dipoles in Electric Double Layer Model - An Analytical Mean-Field Approach. *Electrochimica Acta*, 178:541–545, 2015.

[24] I. J. Ong and J. Newman. Double-Layer Capacitance in a Dual Lithium Ion Insertion Cell. *Journal of The Electrochemical Society*, 146(12):4360–4365, 1999.

[25] S. Santhanagopalan and R. E. White. Quantifying Cell-to-Cell Variations in Lithium Ion Batteries. *International Journal of Electrochemistry*, 2012, 2012. Art. ID 395838.

[26] J. P. Meyers, M. Doyle, R. M. Darling, and J. Newman. The Impedance Response of a Porous Electrode Composed of Intercalation Particles. *Journal of The Electrochemical Society*, 147(8):2930–2940, 2000.

[27] B. S. Haran, B. N. Popov, and R. E. White. Determination of the Hydrogen Diffusion Coefficient in Metal Hydrides by Impedance Spectroscopy. *Journal of Power Sources*, 75(1):56–63, 1998.

[28] M. T. Lawder, P. W. C. Northrop, and V. R. Subramanian. Model-Based SEI Layer Growth and Capacity Fade Analysis for EV and PHEV Batteries and Drive Cycles. *Journal of The Electrochemical Society*, 161(14):A2099–A2108, 2014.

[29] M. Rashid and A. Gupta. Mathematical Model for Combined Effect of SEI Formation and Gas Evolution in Li-Ion Batteries. *ECS Electrochemistry Letters*, 3(10):2014–2017, 2014.

[30] A. M. Colclasure, K. A. Smith, and R. J. Kee. Modeling Detailed Chemistry and Transport for Solid-Electrolyte-Interface (SEI) Films in Li-Ion Batteries. *Electrochimica Acta*, 58(1):33–43, 2011.

[31] E. Prada, D. Di Domenico, Y. Creff, J. Bernard, V. Sauvant-Moynot, and F. Huet. Physics-Based Modelling of $LiFePO_4$-Graphite Li-Ion Batteries for Power and Capacity Fade Predictions: Application to Calendar Aging of PHEV and EV. *2012 IEEE Vehicle Power and Propulsion Conference, VPPC 2012*, pages 301–308, 2012.

[32] G. Sikha, B. N. Popov, and R. E. White. Effect of Porosity on the Capacity Fade of a Lithium-Ion Battery. *Journal of The Electrochemical Society*, 151(7):A1104–A1114, 2004.

[33] H. J. Ploehn, P. Ramadass, and R. E. White. Solvent Diffusion Model for Aging of Lithium-Ion Battery Cells. *Journal of The Electrochemical Society*, 151(3):A456–A462, 2004.

[34] M. Safari and C. Delacourt. Simulation-Based Analysis of Aging Phenomena in a Commercial Graphite/LiFePO4 Cell. *Journal of The Electrochemical Society*, 158(12):A1436–A1447, 2011.

[35] J. Christensen. Modeling Diffusion-Induced Stress in Li-Ion Cells with Porous Electrodes. *Journal of The Electrochemical Society*, 157(3):A366–A380, 2010.

[36] R. Narayanrao, M. M. Joglekar, and S. Inguva. A Phenomenological Degradation Model for Cyclic Aging of Lithium Ion Cell Materials. *Journal of the Electrochemical Society*, 160(1):A125–A137, 2013.

[37] I. Laresgoiti, S. Käbitz, M. Ecker, and D. U. Sauer. Modeling Mechanical Degradation in Lithium Ion Batteries During Cycling : Solid Electrolyte Interphase Fracture. *Journal of Power Sources*, 300:112–122, 2015.

[38] R. Deshpande, M. Verbrugge, Y.-T. Cheng, J. Wang, and P. Liu. Battery Cycle Life Prediction with Coupled Chemical Degradation and Fatigue Mechanics. *Journal of the Electrochemical Society*, 159(10):A1730–A1738, 2012.

[39] S. Santhanagopalan, Q. Guo, and R. E. White. Parameter Estimation and Model Discrimination for a Lithium-Ion Cell. *Journal of The Electrochemical Society*, 154:A198–A206, 2007.

[40] T. G. Zavalis, M. Klett, M. H. Kjell, M. Behm, R. W. Lindström, and G. Lindbergh. Aging in Lithium-Ion Batteries: Model and Experimental Investigation of Harvested $LiFePO_4$ and Mesocarbon Microbead Graphite Electrodes. *Electrochimica Acta*, 110:335–348, 2013.

[41] S. Santhanagopalan, Q. Guo, P. Ramadass, and R. E. White. Review of Models for Predicting the Cycling Performance of Lithium-Ion Batteries. *Journal of Power Sources*, 156(2):620–628, 2006.

[42] E. Prada, D. Di. Domenico, Y. Creff, J. Bernard, and F. Huet. Simplified Electrochemical and Thermal Model of $LiFePO_4$-Graphite Li-Ion Batteries for Fast Charge Applications. *Journal of the Electrochemical Society*, 159(9):1508–1519, 2012.

[43] D. Di Domenico, A. Stefanopoulou, and G. Fiengo. Lithium-Ion Battery State of Charge and Critical Surface Charge Estimation Using an Extended Kalman Filter. *Journal of dynamic systems, measurement, and control*, 132(6):061302, 2010.

[44] D. R. Baker and M. W. Verbrugge. Temperature and Current Distribution in Thin-Film Batteries. *Journal of The Electrochemical Society*, 146(7):2413–2424, 1999.

[45] M. Guo, G. Sikha, and R. E. White. Single-Particle Model for a Lithium-Ion Cell: Thermal Behavior. *Journal of The Electrochemical Society*, 158(2):A122–A132, 2011.

[46] M. B. Pinson and M. Z. Bazant. Theory of SEI Formation in Rechargeable Batteries: Capacity Fade, Accelerated Aging and Lifetime Prediction. *Journal of the Electrochemical Society*, 160(2):A243–A250, 2013.

[47] P. W. C. Northrop, V. Ramadesigan, S. De, and V. R. Subramanian. Coordinate Transformation, Orthogonal Collocation, Model Reformulation and Simulation of Electrochemical-Thermal Behavior of Lithium-Ion Battery Stacks. *Journal of The Electrochemical Society*, 158(12):A1461–A1477, 2011.

[48] M. Xu, Z. Zhang, X. Wang, L. Jia, and L. Yang. A Pseudo Three-Dimensional Electrochemical-Thermal Model of a Prismatic $LiFePO_4$ Battery During Discharge Process. *Energy*, 80:303–317, 2015.

[49] M. Guo and R. E. White. A Distributed Thermal Model for a Li-Ion Electrode Plate Pair. *Journal of Power Sources*, 221:334–344, 2013.

[50] J. Li, Y. Cheng, M. Jia, Y. Tang, Y. Lin, Z. Zhang, and Y. Liu. An Electrochemical-Thermal Model Based on Dynamic Responses for Lithium Iron Phosphate Battery. *Journal of Power Sources*, 255:130–143, 2014.

[51] C. Fink and B. Kaltenegger. Electrothermal and Electrochemical Modeling of Lithium-ion Batteries: 3D Simulation with Experimental Validation. *ECS Transactions*, 61(27):105–124, 2014.

[52] R. E. Garcia, Y.-M. Chiang, W. Craig Carter, P. Limthongkul, and C. M. Bishop. Microstructural Modeling and Design of Rechargeable Lithium-Ion Batteries. *Journal of The Electrochemical Society*, 152:A255–A263, 2005.

[53] R. E. García and Y.-M. Chiang. Spatially Resolved Modeling of Microstructurally Complex. *Journal of The Electrochemical Society*, 154(9):856–864, 2007.

[54] A. Latz and J. Zausch. Multiscale Modeling of Lithium Ion Batteries : Thermal Aspects. *Beilstein journal of nanotechnology*, pages 987–1007, 2015.

[55] T. Hutzenlaub, S. Thiele, N. Paust, R. Spotnitz, R. Zengerle, and C. Walchshofer. Three-Dimensional Electrochemical Li-Ion Battery Modelling

Featuring a Focused Ion-Beam/Scanning Electron Microscopy Based Three-Phase Reconstruction of a $LiCoO_2$ Cathode. *Electrochimica Acta*, 115:131–139, 2014.

[56] G. B. Less, J. H. Seo, S. Han, A. M. Sastry, J. Zausch, A. Latz, S. Schmidt, C. Wieser, D. Kehrwald, and S. Fell. Micro-Scale Modeling of Li-Ion Batteries: Parameterization and Validation. *Journal of The Electrochemical Society*, 159(6):A697–A704, 2012.

[57] A. G. Kashkooli, S. Farhad, D. U. Lee, K. Feng, S. Litster, S. K. Babu, L. Zhu, and Z. Chen. Multiscale Modeling of Lithium-Ion Battery Electrodes Based on Nano-Scale X-Ray Computed Tomography. *Journal of Power Sources*, 307:496–509, 2016.

[58] W. Du, N. Xue, W. Shyy, and J. R. R. A. Martins. A Surrogate-Based Multi-Scale Model for Mass Transport and Electrochemical Kinetics in Lithium-Ion Battery Electrodes. *Journal of the Electrochemical Society*, 161(8):E3086–E3096, 2014.

[59] D. E. Stephenson, E. M. Hartman, J. N. Harb, and D. R. Wheeler. Modeling of Particle-Particle Interactions in Porous Cathodes for Lithium-Ion Batteries. *Journal of The Electrochemical Society*, 154:A1146–A1155, 2007.

[60] F. Röder, S. Sonntag, D. Schröder, and U. Krewer. Simulating the Impact of Particle Size Distribution on the Performance of Graphite Electrodes in Lithium-Ion Batteries. *Energy Technology*, 4:1588–1597, 2016.

[61] M. M. Majdabadi, S. Farhad, M. Farkhondeh, R. A. Fraser, and M. Fowler. Simplified Electrochemical Multi-Particle Model for $LiFePO_4$ Cathodes in Lithium-Ion Batteries. *Journal of Power Sources*, 275:633–643, 2015.

[62] M. Farkhondeh and C. Delacourt. Mathematical Modeling of Commercial $LiFePO_4$ Electrodes Based on Variable Solid-State Diffusivity. *Journal of The Electrochemical Society*, 159(2):A177–A192, 2012.

[63] M. Farkhondeh, M. Safari, M. Pritzker, M. Fowler, T. Han, J. Wang, and C. Delacourt. Full-Range Simulation of a Commercial $LiFePO_4$ Electrode Accounting for Bulk and Surface Effects: A Comparative Analysis. *Journal of the Electrochemical Society*, 161(3):A201–A212, 2013.

[64] M A. Quiroga, K.-H. Xue, T.-K Nguyen, M. Tułodziecki, H. Huang, and A. A. Franco. A Multiscale Model of Electrochemical Double Layers in Energy Conversion and Storage Devices. *Journal of The Electrochemical Society*, 161(8):E3302–E3310, 2014.

[65] H. Wang and L. Pilon. Mesoscale Modeling of Electric Double Layer Capacitors with Three-Dimensional Ordered Structures. *Journal of Power Sources*, 221:252–260, 2013.

[66] G. Blanquer, Y. Yin, M. A. Quiroga, and A. A Franco. Modeling Investigation of the Local Electrochemistry in Lithium-O_2 Batteries: A Kinetic Monte Carlo Approach. *Journal of The Electrochemical Society*, 163(3):329–337, 2016.

[67] J. Yu, M. L. Sushko, S. Kerisit, K. M. Rosso, and J. Liu. Kinetic Monte Carlo Study of Ambipolar Lithium-Ion and Electron-Polaron Diffusion Into Nanostructured TiO_2. *Journal of Physical Chemistry Letters*, 3:2076–2081, 2012.

[68] A. Van der Ven and G Ceder. Lithium Diffusion in Layered Li_x CoO_2. *Electrochemical and Solid-State Letters*, 3(7):301–304, 2000.

[69] J. Bhattacharya and A. Van Der Ven. First-Principles Study of Competing Mechanisms of Nondilute Li Diffusion in Spinel Li_xTiS_2. *Physical Review B*, 83(144302):1–9, 2011.

[70] K. Persson, Y. Hinuma, Y. Meng, A. Van der Ven, and G. Ceder. Thermodynamic and Kinetic Properties of the Li-Graphite System from First-Principles Calculations. *Physical Review B*, 82(12):1–9, 2010.

[71] E. D. Cubuk, W. L. Wang, K. Zhao, J. J. Vlassak, Z. Suo, and E. Kaxiras. Morphological Evolution of Si Nanowires upon Lithiation: A First-Principles Multiscale Model. *Nano Letters*, 13:2011–2015, 2013.

[72] R. N. Methekar, P. W. C. Northrop, K. Chen, R. D. Braatz, and V. R. Subramanian. Kinetic Monte Carlo Simulation of Surface Heterogeneity in Graphite Anodes for Lithium-Ion Batteries: Passive Layer Formation. *Journal of The Electrochemical Society*, 158(4):A363–A370, 2011.

[73] G. Ramos-Sanchez, F. A. Soto, J. M. Martinez de la Hoz, Z. Liu, P. P. Mukherjee, F. El-Mellouhi, J. M. Seminario, and P. B. Balbuena. Computational Studies of Interfacial Reactions at Anode Materials: Initial Stages of the Solid-Electrolyte-Interphase Layer Formation. *Journal of Electrochemical Energy Conversion and Storage*, 13(3):031002–1–031002–10, 2016.

[74] S.-P. Kim, A. C. T. van Duin, and V. B. Shenoy. Effect of Electrolytes on the Structure and Evolution of the Solid Electrolyte Interphase (SEI) in Li-Ion Batteries: A Molecular Dynamics Study. *Journal of Power Sources*, 196(20):8590–8597, 2011.

[75] J. Vatamanu, O. Borodin, and G. D. Smith. Molecular Dynamics Simulation Studies of the Structure of a Mixed Carbonate/$LiPF_6$ Electrolyte Near Graphite

Surface as a Function of Electrode Potential. *Journal of Physical Chemistry*, 116(1):1114–1121, 2012.

[76] P. Ganesh, P. R. C. Kent, and D. Jiang. Solid-electrolyte Interphase Formation and Electrolyte Reduction at Li-Ion Battery Graphite Anodes: Insights from First-Principles Molecular Dynamics. *Journal of Physical Chemistry C*, 116(46):24476–24481, 2012.

[77] K. Leung and J. L. Budzien. Ab Initio Molecular Dynamics Simulations of the Initial Stages of Solid-Electrolyte Interphase Formation on Lithium-Ion Battery Graphitic Anodes. *Physical chemistry chemical physics : PCCP*, 12(25):6583–6586, 2010.

[78] D. Bedrov, G. D. Smith, and A. C. T. Van Duin. Reactions of Singly-Reduced Ethylene Carbonate in Lithium Battery Electrolytes: A Molecular Dynamics Simulation Study Using the ReaxFF. *Journal of Physical Chemistry A*, 116(11):2978–2985, 2012.

[79] J. Yang and J. S. Tse. Li Ion Diffusion Mechanisms in $LiFePO_4$: An Ab Initio Molecular Dynamics Study. *The journal of physical chemistry. A*, 115:13045–13049, 2011.

[80] Y. Wang, S. Nakamura, M. Ue, and P. B. Balbuena. Theoretical Studies to Understand Surface Chemistry on Carbon Anodes for Lithium-Ion Batteries: Reduction Mechanisms of Ethylene Carbonate. *Journal of the American Chemical Society*, 123(47):11708–11718, 2001.

[81] F. Zhou, M. Cococcioni, C. A. Marianetti, D. Morgan, and G. Ceder. First-Principles Prediction of Redox Potentials in Transition-Metal Compounds with LDA + U. *Physical Review B*, 70:1–8, 2004.

[82] Y. C. Chen, C. Y. Ouyang, L. J. Song, and Z. L. Sun. Electrical and Lithium Ion Dynamics in Three Main Components of Solid Electrolyte Interphase from Density Functional Theory Study. *The Journal of Physical Chemistry C*, 115(14):7044–7049, 2011.

[83] S. Shi, Y. Qi, H. Li, and L. G. Hector. Defect Thermodynamics and Diffusion Mechanisms in Li_2CO_3 and Implications for the Solid Electrolyte Interphase in Li-Ion Batteries. *Journal of Physical Chemistry C*, 117(17):8579–8593, 2013.

[84] S. Shi, P. Lu, Z. Liu, Y. Qi, L. G. Hector, H. Li, and S. J. Harris. Direct Calculation of Li-ion Transport in the Solid Electrolyte Interphase. *Journal of the American Chemical Society*, 134(37):15476–15487, 2012.

[85] R. Malik, D. Burch, M. Bazant, and G. Ceder. Particle Size Dependence of the Ionic Diffusivity. *Nano Letters*, pages 4123–4127, 2010.

[86] Z. Zheng, R. M. Stephens, R. D. Braatz, R. C. Alkire, and L. R. Petzold. A Hybrid Multiscale Kinetic Monte Carlo Method for Simulation of Copper Electrodeposition . *Journal of Computational Physics*, 227(10):5184–5199, 2008.

[87] M. A. Quiroga, K. Malek, and A. A. Franco. A Multiparadigm Modeling Investigation of Membrane Chemical Degradation in PEM Fuel Cells. *Journal of The Electrochemical Society*, 163(2):F59–F70, 2016.

[88] S. Pal and D. P. Landau. Monte Carlo Simulation and Dynamic Scaling of Surfaces in MBE Growth. *Physical Review B*, 49(15):10597–10606, 1994.

[89] F. Röder, R. D. Braatz, and U. Krewer. Direct Coupling of Continuum and Kinetic Monte Carlo Models for Multiscale Simulation of Electrochemical Systems. *Computers & Chemical Engineering*, 121:722–735, 2019.

[90] M. Salciccioli, M. Stamatakis, S. Caratzoulas, and D. G. Vlachos. A Review of Multiscale Modeling of Metal-Catalyzed Reactions: Mechanism Development for Complexity and Emergent Behavior. *Chemical Engineering Science*, 66(19):4319–4355, 2011.

[91] K. F. Kalz, R. Kraehnert, M. Dvoyashkin, R. Dittmeyer, R. Gläser, U. Krewer, K. Reuter, and J.-D. Grunwaldt. Future Challenges in Heterogeneous Catalysis: Understanding Catalysts under Dynamic Reaction Conditions. *ChemCatChem*, 9:1–14, 2016.

[92] B. Andreaus and M. Eikerling. Active Site Model for CO Adlayer Electrooxidation on Nanoparticle Catalysts. *Journal of Electroanalytical Chemistry*, 607:121–132, 2007.

[93] T. O. Drews, R. D. Braatz, and R. C. Alkire. Coarse-Grained Kinetic Monte Carlo Simulation of Copper Electrodeposition with Additives. *International Journal for Multiscale Computational Engineering*, 2(2):313–327, 2004.

[94] R. Pornprasertsuk, J. Cheng, H. Huang, and F. B. Prinz. Electrochemical Impedance Analysis of Solid Oxide Fuel Cell Electrolyte Using Kinetic Monte Carlo Technique. *Solid State Ionics*, 178(3–4):195–205, 2007.

[95] T. Jahnke, G. Futter, A. Latz, T. Malkow, G. Papakonstantinou, G. Tsotridis, P. Schott, M. Gérard, M. Quinaud, M. Quiroga, A. A. Franco, K. Malek, F. Calle-Vallejo, R. Ferreira De Morais, T. Kerber, P. Sautet, D. Loffreda, S. Strahl, M. Serra, P. Polverino, C. Pianese, M. Mayur, W. G. Bessler, and

C. Kompis. Performance and Degradation of Proton Exchange Membrane Fuel Cells: State of the Art in Modeling from Atomistic to System Scale. *Journal of Power Sources*, 304:207–233, 2016.

[96] L. A. Ricardez-Sandoval. Current Challenges in the Design and Control of Multiscale Systems. *The Canadian Journal of Chemical Engineering*, 89(December):1324–1341, 2011.

[97] L. Madec, L. Falk, and E. Plasari. Simulatuion of Agglomeration Reactors Via a Coupled CFD/Direct Monte-Claro Method. *Chemical Engineering Science*, 56:1731–1736, 2001.

[98] S. Matera, M. Maestri, A. Cuoci, and K. Reuter. Predictive-quality surface reaction chemistry in real reactor models: Integrating first-principles kinetic monte carlo simulations into computational fluid dynamics. *ACS Catalysis*, 4(11):4081–4092, 2014.

[99] M. A. Quiroga and A. A. Franco. A Multi-Paradigm Computational Model of Materials Electrochemical Reactivity for Energy Conversion and Storage. *Journal of the Electrochemical Society*, 162(7):E73–E83, 2015.

[100] F. Röder, R. D. Braatz, and U. Krewer. Multi-Scale Modeling of Solid Electrolyte Interface Formation in Lithium-Ion Batteries. *Computer Aided Chemical Engineering*, 38:157–162, 2016.

[101] D. Vlachos. Multiscale Integration Hybrid Algorithms for Homogeneous–Heterogeneous Reactors. *AIChE Journal*, 43(11):3031–3041, 1997.

[102] R. D. Braatz, R. C. Alkire, E. Seebauer, E. Rusli, R. Gunawan, T. O. Drews, X. Li, and Y. He. Perspectives on the Design and Control of Multiscale Systems. *J. Process Control*, 16(3):193–204, 2006.

[103] E. Rusli, T. O. Drews, and R. D. Braatz. Systems Analysis and Design of Dynamically Coupled Multiscale Reactor Simulation Codes. *Chemical Engineering Science*, 59:5607–5613, 2004.

[104] E. Weinan, B. Engquist, and Z. Huang. Heterogeneous Multiscale Method: A General Methodology for Multiscale Modeling. *Physical Review B*, 67(092101):1–4, 2003.

[105] A. C. To and S. Li. Perfectly Matched Multiscale Simulations. *Physical Review B*, 72(035414):1–4, 2005.

[106] F. Röder, R. D. Braatz, and U. Krewer. Multi-Scale Simulation of Heterogeneous Surface Film Growth Mechanisms in Lithium-Ion Batteries. *Journal of The Electrochemical Society*, 164(11):E3335–E3344, 2017.

[107] M. Winter. The Solid Electrolyte Interphase – The Most Important and the Least Understood Solid Electrolyte in Rechargeable Li Batteries. *Zeitschrift für Physikalische Chemie, International Journal of Research in Physical Chemistry and Chemical Physics*, 223:1395–1406, 2009.

[108] S. Chattopadhyay, A. L. Lipson, H. J. Karmel, J. D. Emery, T. T. Fister, P. A. Fenter, M. C. Hersam, and M. J. Bedzyk. In Situ X-ray Study of the Solid Electrolyte Interphase (SEI) Formation on Graphene as a Model Li-ion Battery Anode. *Chemistry of Materials*, 24(15):3038–3043, 2012.

[109] S. J. An, J. Li, C. Daniel, D. Mohanty, S. Nagpure, and D. L. Wood. The State of Understanding of the Lithium-Ion-Battery Graphite Solid Electrolyte Interphase (SEI) and Its Relationship to Formation Cycling. *Carbon*, 105:52–76, 2016.

[110] M. B. Pinson and M. Z. Bazant. Theory of SEI Formation in Rechargeable Batteries: Capacity Fade, Accelerated Aging and Lifetime Prediction. *Journal of the Electrochemical Society*, 160(2):A243–A250, 2013.

[111] P. Arora, R. E. White, and M. Doyle. Capcity Fade Mechanisms and Side Reactions in Lithium-Ion Batteries. *Journal of The Electrochemical Society*, 145(10):3647–3667, 1998.

[112] P. Verma, P. Maire, and P. Novák. A Review of the Features and Analyses of the Solid Electrolyte Interphase in Li-ion Batteries. *Electrochimica Acta*, 55(22):6332–6341, 2010.

[113] L. Seidl, S. Martens, J. Ma, U. Stimming, and O. Schneider. In-Situ Scanning Tunneling Microscopy Studies of the SEI Formation on Graphite Electrodes for Li^+-Ion Batteries. *Nanoscale*, 1(3):14004–14014, 2016.

[114] K. Leung and C. M. Tenney. Toward First Principles Prediction of Voltage Dependences of Electrolyte/Electrolyte Interfacial Processes in Lithium-Ion Batteries. *Journal of Physical Chemistry C*, 117:24224–24235, 2013.

[115] E. Peled. Advanced Model for Solid Electrolyte Interphase Electrodes in Liquid and Polymer Electrolytes. *Journal of The Electrochemical Society*, 144(8):L208–L210, 1997.

[116] F. G. Helferich. *Kinetics of Homogeneous Multistep Reactions*, volume 38 of *Comprehensive Chemical Kinetics*. Elsevier Science Ltd, 2001.

[117] A. M. Colclasure and R. J. Kee. Thermodynamically Consistent Modeling of Elementary Electrochemistry in Lithium-Ion Batteries. *Electrochimica Acta*, 55(28):8960–8973, 2010.

[118] J. D. Weeks and G. H. Gilmer. Dynamics of Crystal Growth. *Advances In Chemical Physics*, 40:157–228, 2007.

[119] U Burghaus. *A Practical Guide to Kinetic Monte Carlo Simulations and Classical Molecular Dynamics Simulations : An Example Book*. Nova Science Publ., New York, 2006.

[120] P. Lu and S. J. Harris. Lithium Transport Within the Solid Electrolyte Interphase. *Electrochemistry Communications*, 13(10):1035–1037, 2011.

[121] D. Aurbach. A Comparative Study of Synthetic Graphite and Li Electrodes in Electrolyte Solutions Based on Ethylene Carbonate-Dimethyl Carbonate Mixtures. *Journal of The Electrochemical Society*, 143(12):3809–3820, 1996.

[122] Y. Ein-Eli, B. Markovsky, D. Aurbach, Y. Carmeli, H. Yamin, and S. Luski. Dependence of the Performance of Li-C Intercalation Anodes for Li-ion Secondary Batteries on the Electrolyte Solution Composition. *Electrochimica Acta*, 39(17):2559–2569, 1994.

[123] A. M. Andersson and K. Edstro. Chemical Composition and Morphology of the Elevated Temperature SEI on Graphite. *Journal of The Electrochemical Society*, 148(10):A1100–A1109, 2001.

[124] M. A. Kiani, M. F. Mousavi, and M. S. Rahmanifar. Synthesis of Nano- and Micro-Particles of $LiMn_2O_4$: Electrochemical Investigation and Assessment as a Cathode in Li Battery. *International Journal of Electrochemical Science*, 6(7):2581–2595, 2011.

[125] K. Xu, Y. Lam, S. S. Zhang, T. R. Jow, and T. B. Curtis. Solvation Sheath of Li^+ in Nonaqueous Electrolytes and Its Implication of Graphite/Electrolyte Interface Chemistry. *Journal of Physical Chemistry C*, 111(20):7411–7421, 2007.

[126] V. A. Agubra and J. W. Fergus. The Formation and Stability of the Solid Electrolyte Interface on the Graphite Anode. *Journal of Power Sources*, 268:153–162, 2014.

[127] W. Märkle, C.-Y. Lu, and P. Novák. Morphology of the Solid Electrolyte Interphase on Graphite in Dependency on the Formation Current. *Journal of The Electrochemical Society*, 158(12):A1478–A1482, 2011.

[128] F. German, A. Hintennach, A. LaCroix, D. Thiemig, S. Oswald, F. Scheiba, M. J. Hoffmann, and H. Ehrenberg. Influence of Temperature and Upper Cut-Off Voltage on the Formation of Lithium-Ion Cells. *Journal of Power Sources*, 264:100–107, 2014.

[129] B. K. Antonopoulos, F. Maglia, F. Schmidt-Stein, J. P. Schmidt, and H. E. Hoster. Formation of the Solid Electrolyte Interphase at Constant Potentials: a Model Study on Highly Oriented Pyrolytic Graphite. *Batteries & Supercaps*, 1:1–13, 2018.

[130] E. Peled, D. Bar Tow, A. Merson, A. Gladkich, L. Burstein, and D. Golodnitsky. Composition, Depth Profiles and Lateral Distribution of Materials in the SEI built on HOPG-TOF SIMS and XPS Studies. *Journal of Power Sources*, 97-98:52–57, 2001. Proceedings of the 10th International Meeting on Lithium Batteries.

[131] S. Bertolini and P. B. Balbuena. Buildup of the Solid Electrolyte Interphase on Lithium-Metal Anodes: Reactive Molecular Dynamics Study. *The Journal of Physical Chemistry C*, 122(20):10783–10791, 2018.

[132] S. J. Harris and P. Lu. Effects of Inhomogeneities – Nanoscale to Mesoscale – on the Durability of Li-Ion Batteries. *The Journal of Physical Chemistry C*, 117:6481–6492, 2013.

[133] T. C. Bach, S. F. Schuster, E. Fleder, J. Müller, M. J. Brand, H. Lorrmann, A. Jossen, and G. Sextl. Nonlinear Aging of Cylindrical Lithium-Ion Cells Linked to Heterogeneous Compression. *Journal of Energy Storage*, 5:212–223, 2016.

[134] N. Legrand, S. Raël, B. Knosp, M. Hinaje, P. Desprez, and F. Lapicque. Including Double-Layer Capacitance in Lithium-Ion Battery Mathematical Models. *Journal of Power Sources*, 251:370–378, 2014.

[135] Marco Schleutker, Jochen Bahner, Chih-Long Tsai, Detlef Stolten, and Carsten Korte. On the interfacial charge transfer between solid and liquid li+ electrolytes. *Phys. Chem. Chem. Phys.*, 19:26596–26605, 2017.

[136] E.J. Plichta and W.K. Behl. A Low-Temperature Electrolyte for Lithium and Lithium-Ion Batteries. *Journal of Power Sources*, 88(2):192–196, 2000.

[137] M. Doyle, J. Newman, A. S. Gozdz, C. N. Schmutz, and J.-M. Tarascon. Comparison of Modeling Predictions with Experimental Data from Plastic Lithium Ion Cells. *Journal of the Electrochemical Society*, 143(6):1890–1903, 1996.

[138] Fabian Single, Arnulf Latz, and Birger Horstmann. Identifying the Mechanism of Continued Growth of the Solid–Electrolyte Interphase. *ChemSusChem*, 11:1950–1955, 2018.

[139] Maureen Tang, Sida Lu, and John Newman. Experimental and Theoretical Investigation of Solid-Electrolyte-Interphase Formation Mechanisms on Glassy Carbon. *Journal of The Electrochemical Society*, 159(11):A1775–A1785, 2012.

[140] Fabian Single, Birger Horstmann, and Arnulf Latz. Dynamics and Morphology of Solid Electrolyte Interphase (SEI). *Physical Chemistry Chemical Physics*, 18(27):17810–17814, 2016.

[141] Dongjiang Li, Dmitry Danilov, Zhongru Zhang, Huixin Chen, Yong Yang, and Peter HL Notten. Modeling the SEI-formation on Graphite Electrodes in $LiFePO_4$ Batteries. *Journal of The Electrochemical Society*, 162(6):A858–A869, 2015.

[142] A. A. Tahmasbi, T. Kadyk, and M. H. Eikerling. Statistical Physics-Based Model of Solid Electrolyte Interphase Growth in Lithium Ion Batteries. *Journal of The Electrochemical Society*, 164(6):A1307–A1313, 2017.

[143] E. Peled, D. Golodnitsky, C. Menachem, and D. Bar-Tow. An advanced tool for the selection of electrolyte components for rechargeable lithium batteries. *Journal of the electrochemical Society*, 145(10):3482–3486, 1998.

[144] D. Goers, M. E. Spahr, A. Leone, W. M/"arkleb, and P. Novak. The Influence of the Local Current Density on the Electrochemical Exfoliation of Graphite in Lithium-Ion Battery Negative Electrodes. *Electrochimica Acta*, 56(11):3799–3808, 2011.

[145] M.Q. Xu, W.S. Li, X.X. Zuo, J.S. Liu, and X. Xu. Performance Improvement of Lithium Ion Battery using PC as a Solvent Component and BS as an SEI Forming Additive. *Journal of Power Sources*, 174(2):705–710, 2007. 13th International Meeting on Lithium Batteries.

[146] G. T.-K. Fey, Y. G. Chen, and H.-M. Kao. Electrochemical Properties of $LiFePO_4$ Prepared via Ball-Milling. *Journal of Power Sources*, 189(1):169–178, 2009.

[147] T. Drezen, N.-H. Kwon, P. Bowen, I. Teerlinck, M. Isono, and I. Exnar. Effect of Particle Size on $LiMnPO_4$ Cathodes. *Journal of Power Sources*, 174(2):949–953, 2007.

[148] V. Srinivasan and J. Newman. Discharge Model for the Lithium Iron-Phosphate Electrode. *Journal of The Electrochemical Society*, 151(10):A1517–A1529, 2004.

[149] R. Darling and J. Newman. Modeling a Porous Intercalation Electrode with Two Characteristic Particle Sizes. *Journal of The Electrochemical Society*, 144(12):4201–4208, 1997.

[150] D. Ramkrishna. *Population Balances: Theory and Applications to Particulate Systems in Engineering.* Academic Press, 2000.

[151] K. Smith and C.-Y. Wang. Solid-State Diffusion Limitations on Pulse Operation of a Lithium Ion Cell for Hybrid Electric Vehicles. *Journal of Power Sources*, 161(1):628–639, 2006.

[152] Y. Xie, J. Li, and C. Yuan. Multiphysics Modeling of Lithium Ion Battery Capacity Fading Process with Solid-Electrolyte Interphase Growth by Elementary Reaction Kinetics. *Journal of Power Sources*, 248:172–179, 2014.

[153] J. Christensen and J. Newman. A Mathematical Model for the Lithium-Ion Negative Electrode Solid Electrolyte Interphase. *Journal of The Electrochemical Society*, 151(11):A1977–A1988, 2004.

[154] M. Nishizawa. Measurements of Chemical Diffusion Coefficient of Lithium Ion in Graphitized Mesocarbon Microbeads Using a Microelectrode. *Electrochemical and Solid-State Letters*, 1(1):10–12, 1999.

[155] B. Li, R. Guo, K. Li, C. Lu, and L. Ling. Electrochemical Properties of MCMBs as Anode for Lithium-Ion Battery. *Fuel Chemistry Division*, 47(1):187–188, 2002.

[156] H. Buqa, A. Würsig, D. Goers, L. J. Hardwick, M. Holzapfel, P. Novák, F. Krumeich, and M. E. Spahr. Behaviour of Highly Crystalline Graphites in Lithium-Ion Cells with Propylene Carbonate Containing Electrolytes. *Journal of Power Sources*, 146:134–141, 2005.

[157] M. E. Spahr, D. Goers, W. Märkle, J. Dentzer, A. Würsig, H. Buqa, C. Vix-Guterl, and P. Novák. Overpotentials and Solid Electrolyte Interphase Formation at Porous Graphite Electrodes in Mixed Ethylene Carbonate-Propylene Carbonate Electrolyte Systems. *Electrochimica Acta*, 55(28):8928–8937, 2010.

[158] B. Michalak, H. Sommer, D. Mannes, A. Kaestner, T. Brezesinski, and J. Janek. Gas Evolution in Operating Lithium-Ion Batteries Studied In Situ by Neutron Imaging. *Scientific Reports*, 5(15627):1–9, 2015.

[159] S. G. Rinaldo, P. Urchaga, J. Hu, W. Lee, J. Stumper, C. Rice, and M. Eikerling. Theoretical Analysis of Electrochemical Surface-Area Loss in Supported Nanoparticle Catalysts. *Physical Chemistry Chemical Physics*, 16(48):26876–26886, 2014.

[160] D. L. Marchisio, R. D. Vigil, and O. Fox, R. Quadrature Method of Moments for Aggregation–Breakage Processes. *Journal of Colloid and Interface Science*, 258(2):322–334, 2003.

[161] U. Krewer, D. Schröder, and C. Weinzierl. Scenario-Based Analysis of Potential and Constraints of Alkaline Electrochemical Cells. *Computer Aided Chemical Engineering*, 33:1237–1242, 2014.

[162] U. Krewer, M. Christov, T. Vidakovic', and K. Sundmacher. Impedance Spectroscopic Analysis of the Electrochemical Methanol Oxidation Kinetics. *Journal of Electroanalytical Chemistry*, 589(1):148–159, 2006.

[163] F. Kubannek and U. Krewer. A Cyclone Flow Cell for Quantitative Analysis of Kinetics at Porous Electrodes by Differential Electrochemical Mass Spectrometry . *Electrochimica Acta*, 210:862–873, 2016.

[164] B. B. Berkes, A. Jozwiuk, H. Sommer, T. Brezesinski, and J. Janek. Simultaneous Acquisition of Differential Electrochemical Mass Spectrometry and Infrared Spectroscopy Data for In Situ Characterization of Gas Evolution Reactions in Lithium-Ion Batteries . *Electrochemistry Communications*, 60:64–69, 2015.